WORDS
CAN CHANGE YOUR
BRAIN

WORDS CAN CHANGE YOUR BRAIN

12 Conversation Strategies to Build Trust, Resolve Conflict, and Increase Intimacy

Andrew Newberg, M.D., and
Mark Robert Waldman

Hudson Street Press

HUDSON STREET PRESS
Published by the Penguin Group
Penguin Group (USA) Inc., 375 Hudson Street, New York, New York 10014, U.S.A. • Penguin Group (Canada), 90 Eglinton Avenue East, Suite 700, Toronto, Ontario, Canada M4P 2Y3 (a division of Pearson Penguin Canada Inc.) • Penguin Books Ltd., 80 Strand, London WC2R 0RL, England • Penguin Ireland, 25 St. Stephen's Green, Dublin 2, Ireland (a division of Penguin Books Ltd.) • Penguin Group (Australia), 250 Camberwell Road, Camberwell, Victoria 3124, Australia (a division of Pearson Australia Group Pty. Ltd.) • Penguin Books India Pvt. Ltd., 11 Community Centre, Panchsheel Park, New Delhi – 110 017, India • Penguin Group (NZ), 67 Apollo Drive, Rosedale, Auckland 0632, New Zealand (a division of Pearson New Zealand Ltd.) • Penguin Books (South Africa) (Pty.) Ltd., 24 Sturdee Avenue, Rosebank, Johannesburg 2196, South Africa

Penguin Books Ltd., Registered Offices: 80 Strand, London WC2R 0RL, England

First published by Hudson Street Press, a member of Penguin Group (USA) Inc.

First Printing, June 2012
10 9 8 7 6 5 4 3 2 1

Illustration credits: Page 49 (left to right): Photograph by M. Zacha. By permission of Morguefile.com; courtesy of Public Domain Clip Art; courtesy of Open Clip Art Library. Page 94 and page 95 (all): Permission granted by Alexander Todorov and the Social Cognition and Social Neuroscience Lab, Princeton University. The software used for face generation is FaceGen 3.1. Oosterhof, N. N., & Todorov, A. (2008). The functional basis of face evaluation. Proceedings of the National Academy of Sciences of the USA, 105, 11087–11092. Page 96: Photograph by Griet Cornille. By permission of Morguefile.com and Griet Cornille. Page 98: © Mark Waldman (This derivative illustration has been digitally enhanced and modified).

REGISTERED TRADEMARK—MARCA REGISTRADA

HUDSON STREET PRESS

CIP data is available.
ISBN 978-1-59463-090-3 (hc.)

Printed in the United States of America
Set in Bembo Std Regular
Designed by Eve L. Kirch

PUBLISHER'S NOTE
While the author has made every effort to provide accurate telephone numbers and Internet addresses at the time of publication, neither the publisher nor the author assumes any responsibility for errors, or for changes that occur after publication. Further, publisher does not have any control over and does not assume any responsibility for author or third-party websites or their content.

This book is printed on acid-free paper. ♾

CONTENTS

Authors' Note vii

Part 1

The Evidence: The Neuroscience of Communication, Consciousness, Cooperation, and Trust

Chapter 1: A New Way to Converse 3

Chapter 2: The Power of Words 23

Chapter 3: The Many Languages of the Brain 39

Chapter 4: The Language of Consciousness 53

Chapter 5: The Language of Cooperation 77

Chapter 6: The Language of Trust 87

Part 2

The Strategies: Developing New Communication Skills

Chapter 7: Inner Values: The Foundation of Conscious Living 103

Chapter 8: Twelve Steps to Intimacy, Cooperation, and Trust 121

Chapter 9: Compassionate Communication: Retraining Your Social Brain 147

Part 3

The Application: Practicing Effective Communication with Others

Chapter 10: Compassionate Communication with Loved Ones 165

Chapter 11: Compassionate Communication in the Workplace 183

Chapter 12: Compassionate Communication with Kids 197

Acknowledgments 209

Appendix A: Compassionate Communication Training: CDs, Mp3s, Workbooks, Webinars, and Workshops 211

Appendix B: Compassionate Communication Training Protocol for Couples 213

Appendix C: Compassionate Communication Research Study 217

Notes 219

Index 253

AUTHORS' NOTE

The communication strategies we've developed and presented in this book have grown out of years of evidence-based research conducted by hundreds of neuroscientists and psychologists throughout the world. This book represents a new model for how the brain creates unique language systems that are designed to convey complex information to others. But neuroscience is often difficult to convey in simple language. Sometimes important information can be left out, and sometimes a hypothesis may read as though it were a fact. Furthermore, when it comes to the brain, a single millimeter of tissue can control many processes in addition to the ones we describe in this book. To counter these problems, and to provide the necessary substantiation for this new approach to communicating effectively with others, we've included extensive endnotes, which will also help guide you if you choose to delve more deeply into the neuroscience of empathy, cooperation, and trust.

PART 1

The Evidence

The Neuroscience of Communication, Consciousness, Cooperation, and Trust

CHAPTER 1

A New Way to Converse

Without language, we would find ourselves living in a state of emotional chaos. Our brain has given us the potential to communicate in extraordinary ways, and the ways we choose to use our words can improve the neural functioning of the brain. In fact, a single word has the power to influence the expression of genes that regulate physical and emotional stress.

If we do not continually exercise the brain's language centers, we cripple our neurological ability to deal with the problems we encounter with each other. Language shapes our behavior, and each word we use is imbued with multitudes of personal meaning. The right words, spoken in the right way, can bring us love, money, and respect, while the wrong words—or even the right words spoken in the wrong way—can lead a country to war. We must carefully orchestrate our speech if we want to achieve our goals and bring our dreams to fruition.

Although we are born with the gift of language, research shows that we are surprisingly unskilled when it comes to communicating with others. We often choose our words without thought, oblivious of the emotional effects they can have on others. We talk more than we need to. We listen poorly, without realizing it, and we often fail to pay attention to the subtle meanings conveyed by facial expressions, body gestures, and the tone and cadence of our voice—elements of communication that are often more important than the words we actually say.

These conversational shortcomings are not caused by poor education. Rather they are largely related to an underdeveloped brain, for the areas that govern social awareness, empathy, and related language skills are not fully operational until we're about thirty years old. Despite this neurological handicap, scientific research shows that anyone—young or old—can exercise the language and social-awareness centers of the brain in ways that will enhance their capacity to communicate more effectively with others.

To date we've identified and documented twelve strategies that will enhance the dynamics of any conversation, even with strangers. They can stimulate deep empathy and trust in the listener's brain, and they can be used to interrupt negative thought patterns that, if left unchecked, can actually damage your brain's emotional-regulation circuits.

The Twelve Strategies of Compassionate Communication

1. Relax
2. Stay present
3. Cultivate inner silence
4. Increase positivity
5. Reflect on your deepest values
6. Access a pleasant memory
7. Observe nonverbal cues
8. Express appreciation
9. Speak warmly
10. Speak slowly
11. Speak briefly
12. Listen deeply

In this book, we'll show you how to use these strategies to rapidly develop deep, long-lasting relationships at home and at work. You'll learn how to interrupt unconscious inner speech that generates anxiety, fear, and doubt. You'll build more intimate relationships in your personal life, and you'll build more successful relationships with your clients, employees, and colleagues. You'll create fun, productive collaborations at work; you'll enhance your management skills; and this will translate into more income and sales.

You'll learn how to recognize when another person is lying, and you'll discover how to use your intuition to know what others are thinking before they even speak. You'll even discover how silence can strengthen the power of your communication skills.

We'll also show you a little secret that will change your facial expression in ways that will inspire trust in others. You can change the rate of your speech to influence how the other person feels, and you'll be able to use your body language to convey more meaning than words can ever capture.

If you practice these strategies for just a few minutes each day, you'll think more clearly, you'll enhance your creativity, and you'll generate more authentic dialogues with others. You can even eliminate conflicts before they begin.

Our brain-scan studies, when combined with the latest research in the fields of language, communication, and mindfulness, demonstrate that these strategies can improve memory and cognition while simultaneously lowering stress, anxiety, and irritability—factors that are known to undermine the effectiveness of any conversation or social interaction. As you practice these strategies on a daily basis, your self-confidence and satisfaction in life will grow in ways that can be measured in the laboratory and felt at home.

We call this strategy "Compassionate Communication," and when you use them in your conversations, something quite surprising occurs:

both of your brains begin to align themselves with each other. This special bond is a phenomenon referred to as "neural resonance," and in this enhanced state of mutual attunement two people can accomplish remarkable things together. Why? Because it eliminates the natural defensiveness that normally exists when people casually converse.

The elements of Compassionate Communication can be combined in different ways to fit different situations, and you can integrate them with other communication approaches, thereby making them more effective. You can use Compassionate Communication with children to help them cope more effectively with interpersonal conflicts, to discuss difficult topics, and even help them achieve higher grades in school. It also helps family members and caregivers converse more effectively with people who are suffering from mental illness or various forms of cognitive decline. Psychotherapists and peer-counseling groups have integrated Compassionate Communication into their practices, and it has been embraced by many spiritual and religious organizations that promote interfaith dialogue and nonviolent communication.

Compassionate Communication in the Workplace

Compassionate Communication was originally developed as a tool to help couples build intimacy and resolve conflicts, and it has found its way into the hallways of hospitals and caregiving facilities, where doctors and nurses use it to improve their interactions with patients and colleagues.

Compassionate Communication has also generated strong interest in the boardrooms of corporate America. It reduces work-related stress, which compromises productivity and eventually leads to burnout, and it has proven to be particularly effective for building stronger and more cooperative teams, for improving communication between upper and lower management, and for improving client and customer support, thus leading to increased sales and company loyalty.

Financial and real estate companies have also embraced Compassionate Communication. Donna Phelan, a vice president and investment officer at a major bank, explains:

> In the fast-paced world of business and financial management, my most important responsibility is *client communication.* Service professionals have a tremendous need to integrate the most effective strategies that exist, and this is particularly true when working in volatile financial markets, where people often experience sensory overload due to the flood of information coming from stock-quote monitors, analysts' research reports, urgent e-mails, and multiple phone lines ringing at once. The principles and techniques of Compassionate Communication provide a mind-set that optimizes the dialogue between clients, advisors, and market strategists. The mind-set asks, what do clients want most? The answer: to be listened to, and to be heard, in the briefest period time, with the greatest accuracy, and in a manner that generates mutual respect and trust. In my profession, business success depends on developing these crucial skills, and we have found that advanced training in Compassionate Communication effectively and quickly accomplishes this need.

In personal relationships, poor listening and speaking skills are major causes of disputes and divorce And in the business world, such weaknesses can drive a company into bankruptcy. Thus, any strategy that can teach a person to speak with clarity, brevity, calmness, kindness, and sincerity will increase interpersonal stability in the workplace and at home. For this reason, Compassionate Communication has been incorporated into a core training module in the Executive MBA Program at Loyola Marymount University in Los Angeles. It enhances teamwork performance and the development of socially responsible corporate values, and it helps to reduce the stress generated by the extraordinary demands placed on students who are also managing thriving businesses. As Chris Manning, a professor of finance and real estate, states, "Compassionate Communication offers a cost-effective way to train individuals to communicate more efficiently and professionally with each other while fostering higher levels of openness, trust and interpersonal rapport."[1] Dr. Manning elaborates:

> As a society, we have become word dependent, unaware that words play only a partial role in the overall communication process in business. More important is the sender's skill in conveying an *intended* message and the receiver's skill at *inferring* what that message will be.[2] These nonverbal messages are imbued with feelings, attitudes, and implied values.[3] The strategies built into Compassionate Communication help students, managers, and business executives to recognize and develop these essential nonverbal cues.

Joan Summers also uses a variation of Compassionate Communication when she interviews job applicants for her insurance company. She begins by asking them what their deepest values are (a key component of Compassionate Communication that we'll address in chapter 7). If the applicant's personal values differ from the values of her company, that person is not hired, because she knows such discrepancies will eventually lead to employee dissatisfaction.

Joan then pays attention to how the person engages in dialogue: Do they make the right kind of eye contact? Do they respond to her questions directly and briefly? Is the tone of their voice warm and gentle? Do they exude positivity about themselves, their skills, and their desire to be part of her team? In essence she is using the components of Compassionate Communication to identify those individuals who have a propensity to communicate effectively with others.

At the Holmes Institute, a theological seminary of the Centers for Spiritual Living, ministerial candidates are trained in Compassionate Communication because it helps them to respond with greater sensitivity to the needs of their congregants. Elementary school teachers are also adopting versions of Compassionate Communication because it helps children develop better coping strategies when conflicts arise on the playground.

Deep Listening

Compassionate Communication puts as much emphasis on listening as it does on speaking. Conscientious listening demands that we train our busy minds to remain focused, not only on what the other person is saying, but also on the nonverbal cues reflected in the speaker's voice, face, and body language. Deep listening also interrupts the inner speech that is constantly produced by the language centers of the brain, a phenomenon we'll explore in chapter 3. When we learn how to step back and observe this chattering mind, a new type of silence is created. This allows us to give greater attention to what the other person is saying and bolsters our capacity to intuit what the other person is feeling, including subtle forms of honesty or deceptiveness that are reflected in the micro-expressions of the face.

As recent brain-scan research shows, the more deeply we listen, the more our brain will mirror the activity in the other person's brain. This is what allows us to truly understand another person and to empathize with their sorrows and joys.

Stress and Transformation: Why Old Brains Resist New Tricks

Throughout this book we'll guide you through different strategies that will change the way you listen, speak, and interact with others, but because they are new, you may find yourself resisting them. This resistance is a natural function of the brain. Once a behavior is learned, it slips into unconscious long-term memory, where it can be brought into action with hardly any conscious effort. Even when we've learned a new behavior that's more effective, the earlier memory and behavior are triggered first.

The human brain needs a tremendous amount of energy to function, and it takes even more energy to build new neural circuits to change the way we normally converse with one another. In fact, every change we make in our lifestyle is perceived by the brain as a stressful event, which is why Compassionate Communication gives special attention to developing strategies that decrease stress.

Stress interferes with the neurological mechanisms that govern language production and perception. When we are stressed, the emotional circuits of the limbic brain become active, and the language circuits in the frontal lobe become less active. Communication studies have shown that stress and tension tighten up the muscles of the face in ways that convey suspicion in the minds of others who are watching us. A relaxed demeanor, on the other hand, conveys openness, confidence, and trustworthiness.

When we are under stress, our tone of voice also changes, taking on a quality of irritability and frustration. This will immediately stimulate a defensive reaction in the listener's brain that will undermine the potential of having a productive dialogue even before the conversation begins.

How do you integrate stress reduction and relaxation into a dialogue, especially when you are in the midst of a busy workday? Here's what John Watkins does at his software development firm. He starts the day by standing in a circle with his six departmental heads. The first minute is spent yawning and stretching, which helps to clear everyone's mind of distracting thoughts and irritations. Next each person is allotted thirty seconds to describe what they are currently working on. If they're encountering any problems, or need assistance, other people in the circle can respond with positive suggestions. But again, they must adhere to the "thirty second" rule, which is a crucial component of Compassionate Communication. No criticisms may be expressed, because a single negative thought can disrupt the collaborative process for the rest of the day.

This may sound like a strange ritual for a multimillion-dollar company, but the results speak for themselves: in less than twenty minutes the team can identify the most essential goals for that day and come up with creative suggestions that can be rapidly evaluated, modified, and implemented.

When John's company was tested by an independent research team, there was—after a year of utilizing this strategy—a significant increase in corporate camaraderie and personal satisfaction, along with measurable decreases in personal anxiety and stress. The number of sick days decreased and company loyalty increased, and this translated into a lower rate of employee turnover. In essence, low stress means greater happiness, and as an important research study recently found when examining more than two thousand business divisions of ten large companies, happy people work harder. They're also more imaginative, creative, and productive.[4]

The Meandering Conversation

Compassionate Communication has a long history. It began in 1992 as an informal experiment that Mark developed with a group of transpersonal psychologists and therapists in Los Angeles. At that time there were only three "rules": relax, speak slowly, and take turns saying whatever comes to mind, without censorship.

The premise was simple: if we could speak from the depth of our beings, rather than in the defensive way we normally relate to others, we might be able to communicate our feelings and desires with more honesty, less anger, and greater sensitivity. Furthermore, if we allow ourselves to speak spontaneously from this inner, deeper self, without imposing a specific agenda on the conversation, the dialogue might become more relevant and meaningful for the individuals involved. We might be able to access deeper emotional truths without fear and thereby generate increased intimacy and trust with others.

When we teach Compassionate Communication to others, we pair people up and guide them through a series of well-tested relaxation techniques. This is followed by several imagination and values-related exercises. Then we tell them to let their conversation flow in any direction it wants to take and to make sure that they respond only to what the other person just said. This strategy enables them to interrupt the inner agendas that most of us unconsciously impose on others when we speak.

By responding only to what the other person just said, both speaker and listener learn how to stay focused on the present moment, and this allows a stronger interpersonal connection to be established. To converse without an agenda may seem counterintuitive—and in business this may sound impractical—especially when there are important issues that need to be addressed. But it isn't. One can open the dialogue by addressing a specific topic, for this will set the tone and direction for the conversation; but once the dialogue begins, you need to give both yourself and the person you are speaking with the opportunity to bring up other issues and concerns that if left unaddressed could compromise the outcome that you desire.

If we don't create a "space" in which these hidden concerns and problems can be shared, then we have failed to communicate effectively. Compassionate Communication creates such a space by limiting the speaker's time and increasing active listening. Consciously encouraging spontaneity in dialogue is also one of the best ways to solve problems because it rapidly generates new ideas and solutions that are unlikely to emerge in more constrained forms of dialogue. In business this is called brainstorming. From a neurological perspective, it taps into the creativity that our human frontal lobes are famous for,[5] and which some scientists like to call "cognition without control."[6]

In situations where you are attempting to establish intimacy, following a preconceived agenda can feel like cold manipulation to the other person. The same holds true when talking to clients and colleagues. They too need to be heard, and so a balance must be struck between having an agenda and following the flow of a moment-to-moment exchange. This "flow" experience is a core element of Compassionate Communication, and research shows that it encourages optimal work capacity, with the greatest potential for creativity, and with least amount of effort and conscious control.[7]

In order to give an individual an experiential sense of the power of moment-to-moment spontaneity, we developed a specific training protocol: a twenty-minute scripted exercise in which two people sit down and practice the twelve strategies of Compassionate Communication. You will be guided through this exercise in chapter 9, and if you practice it several times, with different people, you'll begin to see how it can transform an ordinary conversation into a remarkable event. The more you practice the training exercise, the easier it will become to integrate Compassionate Communication into conversations in the real world.

A relaxed, meandering conversation turns out to have other benefits as well. For example, it can reduce social anxiety in people who feel uncomfortable when entering new situations. It also allows a person to gain access to deeper levels of unconscious material without becoming overwhelmed by its contents. This component of Compassionate Communication is related to the Freudian psychoanalytic process of free association and the meditation practice known as mindfulness. Both strategies help an individual to remain relaxed and in the present moment, where they can watch the productions of their busy, noisy mind without becoming caught up in a myriad of distracting thoughts.

The Neuroscience of Mindfulness and Compassion

In the 1970s mindfulness practices were introduced to the medical community, and they are now considered one of the most effective ways to reduce stress and improve health. In the 1990s mindfulness began to transform the world of psychotherapy. By remaining deeply relaxed and observant of their feelings and thoughts, patients were able to reduce their anxiety, depression, and irritability. They didn't have to do anything other than to watch themselves with detachment.

As interest in mindfulness continued to grow, teams of neuroscientists began to explore the neurological correlates of this unusual way of thinking. As they observed the brains of hundreds of people while they practiced various forms of relaxation, stress reduction, and meditation, they discovered a common effect. Mindfulness not only increased a person's ability to control destructive emotions, it also improved the cognitive functioning of the brain, especially in areas relating to language and social awareness.

Our own brain-scan research found that the strategies incorporated in mindfulness could strengthen the neural circuits associated with empathy, compassion, and moral decision making, and it even appears to enhance our ability to be more aware of the workings of our own consciousness. As we began to study the latest findings concerning the coevolution of language and the brain, we realized that the principles of mindfulness could be directly applied to our conversations with other people.

Speaking Briefly and the "Thirty Second" Rule

The neuroscience of language, consciousness, and communication raises many fundamental questions, the answers to which consistently defy definition. For example: When we speak, where do our words come from? Our brain or our mind? And what do we mean by mind? Is it purely a production of the brain, or is it something else? The evidence suggests that the mind and the brain are interconnected, but it remains a mystery as to what, or where, that connection is. Indeed, it even appears that the mind has a "mind" of its own, and so does the brain! Similar dilemmas arise when we try to study the nature of consciousness. Hypotheses abound, but nobody really seems to know.

However, we do have a few clues that illuminate the relationship between the brain, our thoughts, and the ability to communicate effectively. For example, everyday consciousness seems to be dependent on an area of the frontal lobes where short-term "working memory" is processed. Our brain stores a tremendous amount of information in long-term memory, but when carrying out a task it must select only the pieces of information that relate to that task in a meaningful and appropriate way.

How much information can our conscious mind hold in its working memory? About four "chunks," and it can hold them only for thirty seconds or less (we'll explain this in more detail later). This tiny bit of information, contained in a tiny window of time, is what we use to communicate our needs to others. This evidence convinced us to modify Compassionate Communication in a fundamental way: when conversing with others, we realized, we should limit ourselves whenever possible to speaking for no more than twenty or thirty seconds. Even a single sentence can contain more than four chunks of information.

Most people say, "But I need time to explain!" That may be true, but if you talk for several minutes, the other person's brain will only recall a fraction of what you've said, and it might not be the part you wanted to convey. The solution? Brevity followed by intense listening to make sure that the other person has grasped the key points of what you said. If they have, great! You can say another sentence. If not, why move on? If the other person hasn't understood you, what good will it do?

In business, time is money, so brevity is a highly valued trait. In fact, some executives insist that important questions and statements be written down on an index card. Once condensed to fit the card, the most important information can be conveyed in the briefest period of time. It's also a great brain-training exercise. The act of writing down a thought forces us to formulate our message in a meaningful, concise, and accurate way.

When we limit ourselves to speaking for only thirty seconds, the brain quickly adapts by filtering out irrelevant information. There's another advantage to speaking briefly: it limits our ability to express negative emotions.

The Problem of Negativity

Extreme brevity keeps the emotional centers of the brain from sabotaging a conversation. Anger is averted before it begins, and, as we will emphasize throughout this book, anger rarely works. Neuroscience supports this premise, but this discovery contradicts the popular belief that people need to express their feelings of frustration to effectively process anger. If you don't, some therapists believe you're not being honest or true to yourself.

Yet the moment a person expresses even the slightest degree of negativity, it increases negativity in both the speaker's and listener's brains. Instead of getting rid of anger, we increase it, and this can, over time, cause irreparable damage, not only to relationships, but to the brain as well. It can interfere with memory storage and cognitive accuracy, and it can disrupt your ability to properly evaluate and respond to social situations.[8] It interferes with making rational decisions,[9] and you're more likely to feel prejudice toward others.[10] What makes anger particularly dangerous is that it blinds you even to the fact that you're angry; thus it gives you a false sense of certainty, confidence, and optimism.[11]

Expressing anger is destructive, but this does not mean that we should completely repress negative feelings. That too can be quite damaging, because unconscious anger—and the constant flow of stress hormones and neurochemicals it releases—can literally eat you alive, damaging the emotional-regulation centers of the brain.

Research shows that the best way to deal with negativity is to observe it inwardly, without reaction and without judgment. The next step is to consciously reframe each negative feeling and thought by shaping it into a positive, compassionate, and solution-based direction. As the esteemed psychologist Barbara Fredrickson has demonstrated, it's important to generate a minimum of three to five positive thoughts in response to every negative reaction you have. When you do, your work will thrive and your personal relationships will blossom.[12] If you don't, your relationships and work will wither.

There's another way to prevent negativity from creeping into the conversation: express frequent comments of appreciation. The more the better, but they need to be heartfelt and genuine. Talk about positive events in your life and avoid complaining about the world. When it comes to positive and negative feelings, the brain responds like an on-off switch: it cannot focus on both at the same time, and as we will explain in the next chapter, negativity is more powerful. That's why we have to maintain the highest positivity ratio we possibly can if we want our work, relationships, and lives to flourish.[13]

Think Before Speaking

As our research evolved, we found that speaking spontaneously, without censorship, could sometimes cause problems for the listener. So we added another rule: before you speak, ask yourself, can the other person hear what I'm about to say without becoming upset? If the answer is no or even maybe, then put that thought aside for a moment, or write it down on a sheet of paper. At a later time, the other person may be more receptive to what you want to say, and in the meantime you'll be able to think about alternative ways of getting your message across.

In business a poorly phrased statement can undermine an important sale or even cost a person their job. But many people fail to realize that the same principle applies to personal and family relationships. Why do we tend to ignore the strategy of thinking before we speak at home? There are many reasons, but one of the most common is tiredness. Exhaustion from a long day of intense work slows down the compassion circuits in the brain. We become impatient, and we lose some of our ability to think clearly. In this state, negative comments can slip out because we simply don't have the energy to turn them off.

Another reason we may not think before we speak is that we grew up in a family with poor communication skills. Illness and aging can also interfere with the neural circuits governing language and emotion, causing us to speak in ways that are difficult for other people to handle.

Of course expressions of frustration and irritability during conversations are unavoidable, but when they happen, you need to do some reparation work. Sometimes a simple apology will suffice, but the best way to handle an emotional blunder is to ask the other person how they were affected. Just showing interest, and being fully present in your blunder, can be enough to reinstate mutual trust and respect. If you stay deeply relaxed during this delicate exchange of words, you'll be able to handle your frustration, or the other person's irritability, with greater diplomacy and tact.

Unlearning How to Speak

Nearly all the research conducted in the fields of communication suggests that we dialogue poorly with one another. And yet most people believe they are effective communicators. How can that be? How can we be oblivious to our own shortcomings? Neuropsychologists have an explanation: "positivity bias."[14] Believing we are better than we actually are turns out to be neurologically enhancing! It gives us confidence and hope in the most difficult situations; without it, we are more likely to give up and fail. Having a positivity bias helps us to maintain emotional stability, and the part of the brain most activated is the anterior cingulate, a key center for generating compassion toward others.[15]

As we'll explain in the next several chapters, the development of our basic language skills tends to culminate around the age of twelve. It's enough to get us through elementary school, but the finer aspects of communication and social awareness are regulated by parts of the brain that don't become fully operative until our late twenties or early thirties.

The metaphor of riding a bicycle comes to mind. We learn how to ride when we are young, but if you want to excel at bicycling, you have to *unlearn* the bad habits you acquired in your earlier years and replace them with more efficient skills. To be an expert bicyclist, you need to delve into the mechanics of balance and motion, and to immerse yourself fully in the *experience* of riding. And you have to practice, practice, practice.

The same applies to communication. We learn the basics in grammar school and high school, but if you want to excel at communicating, you have to unlearn many bad habits and replace them with advanced skills like empathetic listening. You have to study the mechanics of verbal inflection, and you have to learn how to read facial expressions that most people tend to ignore. You have to immerse yourself fully in the experience of speaking and listening, and you have to practice, practice, practice.

To improve our conversational skills, we need to do four things:

1. Recognize the limits of our personal communication styles.
2. Interrupt old, habituated patterns of conversing.
3. Experiment with new communication strategies long enough to build new neural circuits and behaviors.
4. Consciously apply these strategies when we talk with others.

How long does it take to experience the beneficial effects of these new communication strategies? Based on the data we've gathered, less than an hour. We've been able to measure an 11 percent increase in social intimacy and empathy in individuals who practice Compassionate Communication with two or three different people, for ten minutes each. That's an astonishing finding, and so far there are no other communication strategies that have been able to generate the same degree of effectiveness.

A New Science of Communication

In the first part of this book, we'll present the most recent evidence on how the brain processes language, speech, and listening. We'll explain how language builds a unique brain and how trust and cooperation are developed and conveyed to others. We'll take you through each of the twelve strategies of Compassionate Communication and share with you the neuropsychological studies that support them.

Then we'll guide you through a twenty-minute interpersonal exercise that incorporates these strategies in a way that will enhance the communication circuits of your brain. Along the way, you may discover that many of your old notions of conversing with others need to be jettisoned and replaced by new forms of speaking and listening.

When doubt creeps in—which happens whenever we try to change old behaviors—we ask that you try to suspend your current belief systems as you experiment with the exercises in this book. By assuming a "beginner's mind," we can teach our old brain some newer tricks that will deepen our connection to others.

We'll introduce you to several techniques that effectively eliminate doubt, worry, and procrastination, and in the final chapters we'll share with you how different people—lovers, parents, children, therapists, teachers, financiers, entrepreneurs, and business executives—have applied Compassionate Communication to their work and lives.

We suggest that you spend five or ten minutes each day practicing the different components of Compassionate Communication, first with the people you trust the most, and then with other people in your social and business circles. After a few weeks of practice, you should notice a significant difference in how you relate to others and how they respond to you, even though they may be unfamiliar with the principles you are applying. Ask them if they notice any difference in your communication style. They'll probably pause for a few moments and agree, and in that instant you will have successfully introduced Compassionate Communication to them. You'll generate greater empathy and mutual trust simply by using your words more wisely.

CHAPTER 2

The Power of Words

Words can heal or hurt, and it only takes a few seconds to prove this neurological fact. First ask yourself this question: how relaxed or tense do you feel right now? Next become as relaxed as you possibly can. Take three deep breaths and yawn a few times; this is one of the most effective ways to reduce physical, emotional, and neurological stress.

Now stretch your arms above your head, drop them to your side, and shake out your hands. Gently roll your head around to loosen up the muscles in your neck and shoulders, then take three more deep breaths. Check your body: do you feel more relaxed or tense? Now check your mind: do you feel more alert or tired or calm?

You've just engaged in the first strategy of Compassionate Communication. Later we'll explain how these small actions change your brain and promote more effective dialogue with others. But for this experiment, we want you to stay as relaxed as possible so you can notice the subtle emotional shifts that take place when you see the first series of words on the following page. Take another deep breath and bring your awareness into the present moment. When you are ready, turn the page.

No.

NO.

NO.

NO.

NO!

NO!!!

How did you react when saw those words? Did your eyebrows rise? Did your muscles tighten? Did you smile or tense your face?

If you were in an fMRI scanner—a huge doughnut-shaped magnet that can take a video of the neural changes happening in your brain—we would record, in less than a second, a substantial increase of activity in your amygdala and the release of dozens of stress-producing hormones and neurotransmitters. These chemicals immediately interrupt the normal functioning of your brain, especially those that are involved with logic, reason, language processing, and communication.

And the more you stay focused on negative words and thoughts, the more you can actually damage key structures that regulate your memory, feelings, and emotions.[1] You may disrupt your sleep, your appetite, and the way your brain regulates happiness, longevity, and health.

That's how powerful a single negative word or phrase can be. And if you vocalize your negativity, even more stress chemicals will be released, not only in your brain, but in the listener's brain as well. You'll both experience increased anxiety and irritability, and it will generate mutual distrust, thereby undermining the ability to build empathy and cooperation. The same thing happens in your brain when you listen to arguments on the radio or see a violent scene in a movie. The brain, it turns out, doesn't distinguish between fantasies and facts when it perceives a negative event. Instead it assumes that a real danger exists in the world.

Any form of negative rumination—for example, worrying about your financial future or health—will stimulate the release of destructive neurochemicals. And if you are prone to constantly thinking about negative possibilities and persistently ruminating about problems that have occurred in the past, you may ultimately test positive for clinical depression.[2] The same holds true for children: the more negative thoughts they have, the more likely they are to experience emotional turmoil.[3] But if you teach them to think positively, you can turn their lives around.[4]

Negative thinking is also self-perpetuating: the more you are exposed to it—your own or other people's—the more your brain will generate additional negative feelings and thoughts. In fact, the human brain seems to have a capacity to spend more time worrying than that of any other creature on the planet. And if you bring that negativity into your speech, you can pull everyone around you into a downward spiral that may eventually lead to violence. And the more you engage in negative dialogue, at home or at work, the more difficult it becomes to stop.[5]

Fearful Words

Angry words send alarm messages through the brain, and they partially shut down the logic-and-reasoning centers located in the frontal lobes. But what about fearful words—words like "poverty," "sickness," "loneliness," and "death"? These too stimulate many centers of the brain, but they have a different effect from negative words. The fight-or-flight reaction triggered by the amygdala causes us to begin to fantasize about negative outcomes, and the brain then begins to rehearse possible counterstrategies for events that may or may not occur in the future.[6] In other words, we overtax our brains by ruminating on fearful fantasies.

Curiously, we seem to be hardwired to worry—perhaps an artifact of ancient memories carried over from ancestral times when there were countless threats to our survival.[7] However, most of the worries we have today are not about really serious threats. We can learn how to retrain our brain by interrupting these negative thoughts and fears. By redirecting our awareness to setting positive goals and building a strong, optimistic sense of accomplishment, we strengthen the areas in our frontal lobe that suppress our tendency to react to imaginary fears. Not only do we build neural circuits relating to happiness, contentment, and life-satisfaction, we also strengthen specific circuits that enhance our social awareness and our ability to empathize with others. This is the ideal state in which effective communication can prosper.

Interrupting Negative Thoughts and Feelings

Several steps can be taken to interrupt the natural propensity to worry. First ask yourself this question: is the situation *really* a threat to my personal survival? The answer is almost always no. Our imaginative but unrealistic frontal lobes are simply fantasizing about a catastrophic event.

The next step is to reframe a negative thought into a positive one. Instead of worrying about your financial situation—which won't have any effect on your income—think about the ways that you can generate more money and keep your mind focused on the steps you need to take to achieve your financial goals.

The same holds true for personal relationships. For example, if you are the type of person who worries about being rejected or perceived in the wrong way by others, shift your focus to those qualities that you truly admire about yourself. Then, when you talk to others, talk about the things your really love and deeply value. And don't talk about your personal problems or the catastrophes happening in the world; they'll enmesh you in feelings of self-doubt and insecurity.

The faster we can interrupt the amygdala's reaction to real or imaginary threats, the quicker we can generate a feeling of safety and well-being and extinguish the possibility of forming a permanent negative memory in our brain.[8] By shifting our language from worry to optimism, we maximize our potential to succeed at any realistic goal we truly desire.

Words Shape the Reality We Perceive

Human brains like to ruminate on negative fantasies, and they're also odd in another way: they respond to positive and negative fantasies as if they were real. Moviemakers make use of this phenomenon all the time. When the green three-eyed monster jumps out of the closet, we jump out of our seat. This is what makes nightmares so frightening for children, whose brains have yet to develop clear distinctions between language-based fantasies and reality.

To make matters worse, the more emotional we get, the more real the imaginary thought becomes. But imagination is a two-way street. If you intensely focus on a word like "peace" or "love," the emotional centers in the brain calm down. The outside world hasn't changed at all, but you will still feel more safe and secure. This is the neurological power of positive thinking, and to date it has been supported by hundreds of well-designed studies. In fact, if you simply practice staying relaxed, as we asked you to do at the beginning of this chapter, and repetitively focus on positive words and images, anxiety and depression will decrease and the number of your unconscious negative thoughts will decline.[9]

When doctors and therapists teach patients to reframe negative thoughts and worries into positive affirmations, the communication process improves and the patient regains self-control and confidence.[10] Indeed just seeing a list of positive words for a few seconds will make a highly anxious or depressed person feel better, and people who use more positive words tend have greater control over emotional regulation.[11]

Growing Positivity

Certain positive words—like "peace" or "love"—may actually have the power to alter the expression of genes throughout the brain and body, turning them on and off in ways that lowers the amount of physical and emotional stress we normally experience throughout the day.[12] But these types of words cannot be grasped by a child's immature brain. Young children do not have the neural capacity to think in abstract terms, so the first words they learn are associated with simple, concrete images and actions. Verbs like "run" or "eat" can be easily associated with visual images, but abstract verbs like "love" or "share" demand far more neural activity than a young child's brain can muster.[13]

Even more neural processing is required when it comes to highly abstract concepts like "peace" or "compassion." This is easy to test. See how long it takes you to visualize the word "table." In less than a second, you can grasp its shape and function, and you can see it in your mind's eye. Now think about the word "justice." You'll realize that it takes much longer to identify an image, and most people will envision the well-known picture of a woman holding a scale. Obviously justice is far more complex than that image can impart, which explains why so few people can agree on what this important concept means. Abstract words make greater demands on many areas of the brain,[14] whereas concrete words require less neural activity.

Abstract thoughts may be essential for solving complex problems, but they also distance us from deeper feelings, especially those that are needed to bind us to other people. In fact, some people can become so involved with abstract concepts that they partially lose touch with reality.[15] Love is a perfect example because we can easily project our ideal notions onto a potential partner, thereby blinding ourselves to the other person's faults. Why does it take so many years to discover what love actually is? Neuroscience has an answer: love turns out to be expressed through one of the most complicated and complex circuitries identified in the human brain.[16] Thus the *language* of love may be the most sophisticated communication process of all.

Abstract concepts can also be sources of miscommunication and conflict because we rarely explain to others what these complex terms mean to us. Instead we make the mistake of assuming that other people share the same meanings that we have imposed on our words. They don't. Let's take, for example, the word "God." In our research we queried thousands of people, using a variety of surveys and questionnaires, and discovered that 90 percent of the respondents had definitions that differed significantly from everyone else. Even people who came from the same religious or spiritual background had fundamentally unique perceptions of what this word means. And for the most part, they never realized that the person they were talking to about God had something entirely different in mind.

Our advice: when an important abstract concept comes up in conversation, take a few minutes to explore what it means to each of you. Don't take your words, or the other person's, for granted. When you take the time to converse about important values and beliefs, clarifying terms will help both of you avoid later conflicts and confusion.

The Power of Yes

What about the power of the word "yes?" Using brain-scan technology, we now have a very good idea of what happens when we hear positive words and phrases. What do we see? Not much! Positive words do not connote a threat to our survival, so our brain doesn't need to respond as rapidly as it does to the word "no."[17] This presents a problem because evidence showing that positive thinking is essential for developing healthy relationships and work productivity continues to grow.

Can we train our brain to become more responsive to "yes"? We think so but in an indirect way, through intense, repetitive focusing on positive images, feelings, and beliefs. And it doesn't matter if the positive thinking is grounded in science, business, or theology. In fact, positive irrational beliefs have also been proven to enhance a person's sense of happiness, well-being, and lifetime satisfaction.[18] Positive thinking can help even people who are born with a genetic propensity toward unhappiness to build a better and more optimistic attitude toward life.[19]

In a landmark study that put "positive psychology" on the map, a large group of adults, ranging in age from thirty-five to fifty-four, were asked to write down, each night, three things that went well for them that day, and to provide a brief explanation why. Over the next three months, their degrees of happiness continued to increase, and their feelings of depression continued to decrease, even though they had discontinued the writing experiment.[20] Thus by using language to help us reflect on positive ideas and emotions, we can enhance our overall well-being and improve the functioning of our brain.

Positive words and thoughts propel the motivational centers of the brain into action,[21] and they help us build up resilience when we are faced with the myriad problems of life.[22] According to Sonja Lyubomirsky, one of the world's leading researchers on happiness, if you want to develop lifelong satisfaction, you should regularly engage in positive thinking about yourself, share your happiest events with others, and savor every positive experience in your life. If you use language—your inner dialogues, your conversations with others, your words, your speech—to engage in optimism and positivity, you will find yourself moving in a more life-enhancing direction.

Can positive thinking be taken too far? Yes, especially if you engage in exaggeration. People may begin to distrust you because the overuse of extremely positive words in speech or writing can be read as a signal that you are being deceptive.[23] This happens quite often in business communication and advertising, and it isn't that the public has become more savvy. It's a natural function of your brain, which is specifically designed to look for dishonesty in a person's face or tone of voice. The solution to this communication problem is to be positive but honest. You don't have to oversell yourself, because if you truly believe in the product or service you are offering—if your words *feel* genuine to you—other people will intuit your authenticity from the nonverbal communication cues you give out.

Here are some examples of words that turn prospective friends and customers off: "amazing," "excellent," "fabulous," "fantastic," "incredible," "marvelous," "great," "phenomenal," "splendid," and "wonderful." Ironically, extremely negative words, especially if directed toward an opponent, appear to give the speaker more credibility in the eyes of the listener by casting doubt on the other person. It's just another example of the power of no.

People can become immune to the overuse of strongly positive or negative words.[24] Their awareness and sensitivity decreases, which may explain why chronic complainers are often unaware of their negativity and the emotional damage they are causing.

Words Can Change Your Genes

As mentioned earlier, certain positive words can, if focused on for ten or twenty minutes per day, influence genetic expression in your brain. In a recent study, Herbert Benson's team at Massachusetts General Hospital discovered that the repetition of personally meaningful words can actually turn on stress-reducing genes.[25] But you have to remain in a deeply relaxed state. To help subjects achieve this state, they were taught to use Benson's "relaxation response." It's very easy to do, and we've described a variation of it in the accompanying sidebar.

Turn on Your Genes, Turn off Your Stress

Sit in a comfortable chair and close your eyes. Take ten deep breaths as you relax every muscle in your body. Now repeat to yourself, silently or aloud, a word or short phrase that gives you a feeling of serenity, peacefulness, or joy. Continue for ten to twenty minutes as you slowly breathe through your nose. Whenever a distracting thought or feeling intrudes, notice it without judgment and let it float away as you return to the repetition of your word. When you finish, open your eyes and notice how you feel. After a few weeks of practice, you'll feel more relaxed and alert, less anxious and depressed. You may even find that you lose some of your desire to smoke, drink, or overeat.

Even novices who had never practiced any form of meditation or relaxation strategy were able to alter their genetic expression in eight weeks. Subjects were each given a twenty-minute CD that guided them through exercises involving diaphragmatic breathing, a "body scan" that involves consciously bringing attention to areas of tension in the body, and the repetition of a single word or phrase that generates a sense of peacefulness and well-being. The researchers suggested that similar practices, including various forms of meditation, repetitive prayer, yoga, tai chi, breathing exercises, progressive muscle relaxation, biofeedback, and guided imagery would have similar effects on our genes. And as you will see in chapter 9, our Compassionate Communication training exercise includes a similar relaxation exercise.

How about negative words? There is mounting evidence that strongly negative terms can interrupt the normal expression of genes that regulate one of the most important language centers of the brain, Wernicke's area.[26] This is where we learn how to interpret the meaning of words. Hostile language also appears to disrupt specific genes that are instrumental in the production of neurochemicals that protect us from physiological stress, and if we are exposed to it during childhood, it can undermine our ability to fend off anxiety, depression, and fear. Hearing hostile language has also been shown to lead to negative ruminations, which can likewise damage our brain.

Can Subliminal Words Influence Behavior?

New research demonstrates that subliminal messages can unconsciously influence our thoughts, feelings, and actions. For example, words and phrases repeated at a volume that we can barely perceive can create subtle changes in mood.[27] Negative words stimulate anxiety, and positive words can lower it.[28] But again the studies consistently show that the brain gives more attention to negative words, even when we are not aware that we've heard them.[29] This reinforces our argument that even the subtlest forms of negativity can sour a relationship. We may murmur a complaint under our breath, but our voice and face will give us away.

On the positive side, subliminal messages can be used to motivate us to do better work.[30] And in personal relationships, subliminal erotic words can trigger intimacy-related thoughts. That should come as no surprise. What is surprising is that erotic words appear to improve a person's conflict-resolution strategies![31] In fact, just hearing a beloved's name, even if we are unconscious of it, will stimulate the circuits related to passion, whereas hearing a friend's name will not.[32]

This has powerful ramifications for intimate relationships, because it tells us how important it is to communicate our feelings of love as often as we possibly can. Unfortunately, we often fall into the habit of taking our loved ones for granted, and thus we tend to speak up only when something bothers us.

But subliminal words are not as effective as persuasive messages that are clearly spoken or written. At the University of California, Los Angeles, researchers put subjects in an fMRI scanner and had them read and listen to messages encouraging the use of sunscreen. The more they were exposed to the messages, the more the subjects used sunscreen in the following week, even though there was no encouragement from the researchers to do so.[33] The authors of this study replicated their findings with smokers: all the participants reduced the number of cigarettes they smoked over the next month, and those with the greatest increase in brain activity demonstrated the greatest decline in smoking.[34]

Transforming Reality

By holding a positive and optimistic thought in your mind, you stimulate frontal lobe activity. This area includes specific language centers that connect directly to the motor cortex responsible for moving you into action.[35] And as our research has shown, the longer you concentrate on positive words, the more you begin to affect other areas of the brain. Functions in the parietal lobe start to change, which changes your perception of yourself and the people you interact with. A positive view of yourself will bias you toward seeing the good in others, whereas a negative self-image will incline you toward suspicion and doubt. Over time the structure of your thalamus will also change in response to your conscious words, thoughts, and feelings, and we believe that the thalamic changes affect the way in which you perceive reality.

Let me give you an example. If you repetitiously focus on the word "peace," saying it aloud or silently, you will begin to experience a sense of peacefulness in yourself and in others. The thalamus will respond to this incoming message of peace, and it will relay the information to the rest of the brain. Pleasure chemicals like dopamine will be released, the reward system of your brain will be stimulated, anxieties and doubts will fade away, and your entire body will relax. And if you do these practices consistently over a period of time, your sense of compassion will grow. In fact, some of the most recent studies show that this kind of exercise will increase the thickness of your neocortex and shrink the size of your amygdala, the fight-or-flight mechanism in your brain.

Our own brain-scan research shows that concentrating and meditating on positive thoughts, feelings, and outcomes can be more powerful than any drug in the world, especially when it comes to changing old habits, behaviors, and beliefs. And to the best of our knowledge, the entire process is driven by the language-based processes of the brain.

By changing the way you use language, you change your consciousness, and that, in turn, influences every thought, feeling, and behavior in your life. Over time you may even begin to ameliorate limiting and disturbing memories by talking about them in a relaxed and positive way. When you do this, the old memory is changed and filed away in a slightly different way.[36] The next time it's recalled, it incorporates some of the new positive language that you encoded it with.

Positive refocusing, positive affirmations, acceptance-based awareness exercises, relaxation, hypnosis, and meditation have all been shown to be effective in interrupting negative ruminations and depressive thoughts,[37] so why not include them in your daily routines? By changing your inner language, you can transform the reality in which you live.

Preventing Ruminations

To undermine negative ruminations, Robert Leahy, a clinical professor of psychology at Weill-Cornell Medical School, recommends that you try the following:

1. Ask yourself if your negative thinking has ever helped you in the past. Usually the answer is no.
2. Write down your negative thoughts, and then put the sheet of paper aside. When you look at it later, the problem won't seem as large.
3. Ask yourself if the problem is real or imaginary. Is it part of the present or part of the past. Accept the past and let it go.
4. Instead of focusing on your problem, focus on an immediate goal that you can accomplish.
5. Accept that many problems are unpleasant, difficult, and unfair and that some of them simply can't be solved.
6. Take a break and focus on doing something enjoyable.

The human brain is incredibly creative, and it dreams up positive and negative scenarios all day long. But most of us aren't aware of these forms of mental chatter. And even when you discover them, or point them out in someone else's behavior, they can go on repeating themselves like a well-worn groove in a record. Why? Because repetitious patterns of thinking form strong neural pathways that are highly resistant to change. That's why we have to continually impose new styles of thinking, speaking, and listening to get new neural circuits to form.

That's the power of imagination: it can trap us in a downward spiral of negative thoughts, or we can use it change decades of habituated behaviors that no longer serve us well.

Minding Our Words

The first step is to recognize that we have all kinds of negative thoughts flowing unconsciously through our mind. Then all we have to do is turn our awareness inward and pay close attention to the processes of the busy brain. We don't have to do anything with what we see or hear; we simply observe, without judging them, the moment-to-moment changes in our thoughts and feelings and sensations. This is the formal definition of mindfulness, and it is a very important tool when it comes to changing the way we think and feel.

Try this experiment right now. Close your eyes and see how long you can remain empty-headed before a thought or feeling intervenes. If you are new at this exercise, you might be able to sit in complete inner silence for just five or ten seconds. And even if you are a seasoned practitioner of mindfulness, you'll rarely be able to go for more than thirty seconds before the mental chatter kicks in.

In mindfulness the purpose is not to remain silent but to become aware of the continual shifts of consciousness that are taking place, a consciousness that is primarily language driven and is filled with opinions, beliefs, conjectures, and plans, with an occasional insight or two. By learning how to passively watch all these inner voices, you'll become aware of the other sounds your mind has filtered out.

And then—just when you think you've quieted your mind—a

cacophony of complaints might erupt. For example, you might find yourself thinking, "This is stupid! I've got more important things to do!" In mindfulness you'll note that thought and then you allow it to float away as you bring your attention back to a state of inner silence or to your breath. But it won't be long before another thought or feeling intrudes, like, "My back hurts!"

This inner dialogue never seems to stop, and it doesn't have to. Your task is to simply observe, without judging it. It's a unique form of awareness that makes your frontal lobes light up like the Fourth of July. When this happens, the brain's ability to generate feelings of anxiety, irritability, or stress are suppressed. Thus when you learn how stay in this state of awareness while you work, you'll accomplish more without getting burned out. You'll feel more satisfied with yourself and with your work, and as some of the newest research has found, you'll even act with greater generosity toward others.[38] As one corporate researcher remarked, it will improve the gross national happiness.[39] This is the neuroeconomics of business psychology, and research shows that with mindful observation and alteration of the inner voices of consciousness, corporate collaboration and management improves.[40]

When you add optimistic thinking to this equation, you can actually add two years to your life.[41] That's what the prestigious Mayo Clinic found in a study that followed seven thousand people for more than forty years. So choose your words wisely, because they will influence your happiness, your relationships, and your personal wealth.

CHAPTER 3

The Many Languages of the Brain

When I, Mark, was seven years old, my parents took me to the United Nations. I had no idea what to expect, but I was awed by the variety of languages I heard. Each sound was like a different flavor in an ice cream store.

We sat in the viewer's gallery and were given headphones that were plugged into our chairs. How fun! I could turn the knob and a different voice would come out, in a different language. But I was confused. A man was talking on the main floor of the auditorium, but the voice coming out of the headphones was a woman's.

I didn't make the connection that I was listening to translators. My dad came to the rescue and pointed to a glass-enclosed room in the back of the hall. It was filled with a dozen people simultaneously talking into microphones. He explained that they were translating what the speaker was saying so that everyone else, from different parts of the world, could understand.

Mystery solved. But in hindsight I now compare that experience to the way our brain processes language. The neurons in our brain have many different ways to communicate information to one another.

Some forms are chemical, others are electrical, and there may even be other dimensions of communication that take place on a subatomic level. At any given moment, dozens of neurotransmitters are communicating different types of information to different cells. We have axons communicating to dendrites, glial cells communicating through calcium waves, white matter promoting communication between different areas of gray matter, right and left hemispheres constantly communicating with each other, and there are even discrete forms of neural oscillation that may help synchronize the overall activity of the brain.

Somewhere within this cacophony of neural dialogues, a little bit of consciousness arises, and it is through this tiny window of inner perception that we communicate our feelings and thoughts to others. Even here there are dozens of language styles. There's verbal and nonverbal language. There's the language of emotion and the language of abstract reasoning. There's body language and sign language. And then there are the languages of the arts: music, poetry, painting, dance, sculpture, song, etc. These too are considered to be unique language systems of the brain, and each has to be developed through education and training.[1]

With the help of brain-imaging technology, we are beginning to see how each of these systems works with the others. Sometimes we can even see where a single word or image might be stored. For example, researchers have been able to locate single neurons that can hold enough information to recognize an image of the Eiffel Tower, Bill Clinton, or your grandmother.[2]

Where Does Language Begin?

It's fair to say that language may begin at the moment of conception, when two strands of DNA begin to interact with each other. As embryonic cells divide, they pass on their genetic language codes to other cells. Cells begin to group together into specialized communities, and they use their own systems of language to coordinate their activities. As the organism evolves and becomes more complex, even more complex systems of communication evolve, and different cellular communities take on different roles.

As in a well-designed business, some groups assume a management position, others take on the role of production, and others begin to engineer structural changes that make the organism function more efficiently. Some groups of cells become inventive, others act as regulators, and some just sit around and worry about potential threats. In essence the brain becomes a vast community of different cultures communicating in vastly different ways for the purpose of maintaining the health of the entire system. But if communication breaks down in even the slightest way—because of disease or genetic abnormalities—the survival of the entire organism can be threatened.

At the same time that the brain is orchestrating its world of inner communication, it also has to learn how to communicate effectively with other brains that have grown up in different environments. Thus the next level of training requires that we agree upon a common language that we can speak and write. New neural processes must be developed. We have to learn how to control our vocal cords and facial expressions to pronounce words with clarity, and we also have to develop sophisticated auditory skills to identify the huge variety of sounds that continually bombard our ears. These language skills take decades to develop, which is why children and young adults are so poor at communicating effectively with others.

The Evolution of Speech

Verbal speech turns out to be one of the most advanced and complicated processes of communication. First you have to have the physiology to make sounds and gestures. Gestures are controlled by the most ancient structures in the brain, which is why gesturing is a common form of communication throughout the animal kingdom.

Speech requires a more complex brain, and the structures that support it are located in the neocortex, which literally means "new brain." This thin outermost surface covers the more ancient emotional brain, and it contains many of the executive functions associated with language acquisition, vocal control, and a variety of interpretive functions that allow us to transform sounds into meaningful expressions that can be understood by others. Without these advanced language centers, we couldn't form a concept of ourselves, nor could we use our creativity to consciously change our lives.

Our language centers have another unique ability not found in other animal species: the neurons of the neocortex can grow axons—the communicating ends of the neuronal body—that extend all the way back into the cerebellum and other parts of the brain that control the movements of our body.[3] This gives us remarkable control over our vocal cords, facial expressions, and hand movements—three core elements in our power to be prolific communicators.

Within the animal kingdom, we alone can actually *think* ourselves into developing more refined movements in our fingers, our face, and our voice. According to many researchers, this coevolution of language and the brain has given us the ability to speak with great precision.[4] And the more we speak and write, the more we strengthen the language connections in the brain.

Bird Brains, Human Brains, and Expressing Your Inner Ape

All living organisms communicate in one way or another. But the question remains: are humans superior to animals when it comes to communication? Yes and no. Ants, for example, have ten thousand neurons, only one-millionth of the number in a human brain, and yet they can coordinate social activity more effectively than any society in the world. As a group they're more peaceful, and when attacked, they are far more efficient at waging war. They understand what their societal roles are, and they can be very creative when it comes to building and maintaining their communities. Compared to the communication strategies of ants, human communicational abilities pale.

Primate vocalization turns out to be quite similar to our own,[5] and the same can be said of birds. They too develop sophisticated forms of vocal communication, and some species have evolved neural language networks that are surprisingly similar to areas in the human brain.[6]

So what makes human communication unique? It's not just the quality of our speech but the quantity. We use tens of thousands of facial expressions, body movements, and words, and we can combine them in endless combinations that allow us to express different nuances of meaning and emotion. Even a simple alteration of the rate and rhythm of our speech can change the context of what we say and the way it will be processed in the listener's brain.

What about men and women? Yes, there are significant neurological differences, but despite the plethora of popular books written on the subject there is little evidence to show that one sex communicates better than the other. Except when it comes to talkativeness. Can you guess which sex is more guilty? Men! They also tend to be more assertive with their speech, and women tend to use more positive relational words than men, but the differences are small.[7]

A Finger Can Speak a Thousand Angry Words

Words themselves do not communicate all the essential elements of what we need and want to convey to others. The expressions we make with our faces, the tone we use when we speak, and the gestures we make with our body are also key to communicating effectively. In fact, your brain needs to integrate both the sounds and body movements of the person who is speaking to accurately perceive what is meant.[8] Furthermore, gestures actually help orchestrate the brain's language comprehension centers.[9]

Paul Ekman, the world's foremost expert on human nonverbal communication, has identified more than ten thousand discrete human facial expressions,[10] and it turns out that the neural networks that control language are the same ones we use for gesturing.[11] Gesturing enhances our memory and comprehension skills,[12] and, depending on which hand you use, your gestures may be conveying information that will influence how the listener responds. For example, when researchers at the Max Planck Institute studied the communication styles of American presidential candidates during the final debates of the 2004 and 2008 elections, they made some fascinating discoveries. In right-handed politicians, positive messages were associated with right-hand gestures, while negative messages were conveyed with gestures by the left hand. For left-handed politicians, the findings were reversed.[13]

A recent Stanford University study confirmed this finding: we tend to express positive ideas with our dominant hand and negative ideas with the other hand.[14] But don't try to second-guess someone by looking at their hand movements alone; there's often a mismatch between speech and gestures, especially when a person is trying to communicate something difficult or new.[15] As Ekman points out, facial expressions and body gestures only give us clues about what the person may actually be trying to convey.

Biologists who study the evolution of human speech have demonstrated that spoken language emerges from our use of hand and facial gestures, and a recent neuroimaging study showed that hand gestures and speech originate in the same language-related area of the brain.[16] This overlap between words and gestures appears to be associated with a rare cluster of brain cells called "mirror neurons."[17] The neurons that fire in someone's brain when they make a specific gesture also fire in your brain as you observe them. Many of these mirror neurons are located in the brain's language centers, and they may be crucial for governing our ability to empathize and cooperate with others.[18]

These neuroscientific studies teach us how important it is to pay close attention to the nonverbal messages given to us by others and to train ourselves to communicate more fully by consciously using our facial expressions, tone of voice, and body language. If our words and gestures are incongruent, a form of neural dissonance occurs that will confuse the person who is listening and watching.[19]

Learn How to Speak with Your Body

Here's a simple exercise that will give you an experiential sense of how words and gestures interact in the brain. Say the following sentence out loud and notice whatever thoughts, images, or feelings come to mind:

THE BALL IS ROUND

Now, say it again slowly, but this time cup your hands as if you are holding a large grapefruit. Notice how it changes your imagery and feelings. Once again, say aloud, "The ball is round," but this time make a huge arc with your hands and arms. It should feel very different, and it should even affect the tone of your voice. When you consciously orchestrate your words with your gestures, you rivet the attention of the listener. Comedians are masters of this technique, and without such gestures, the humor behind their words can be lost.

If you want to become more effective at communicating with your facial gestures, Paul Ekman recommends that you stand in front of a mirror and imitate expressions of anger, sadness, and fear. Ekman found that when you make such facial gestures, "You will trigger changes in your physiology, both in your body and your brain."[20] With practice, you'll learn how to identify these disruptive emotions in yourself, before they can derail a conversation with someone else.

Practicing expressions relating to happiness and satisfaction turns out to be a little trickier. You can try it in the mirror, but you'll soon discover that the slightest changes in a smile convey different meanings, ranging from anxiety to contentment. Because the expressions of emotion are often controlled by involuntary muscles, it's much harder fake honesty, love, and trustworthiness. And yet most people, when they do feel compassion and kindness, are not able to fully show these expressions on their face. In chapter 5 we'll explain how you can consciously and deliberately generate these important facial cues when you engage others in conversations requiring trust and rapport.

Each time you enter a conversation with another person, pay attention to the vast number of nonverbal expressions being made. Then try to coordinate your hand gestures, gaze, and body posture with the other person. When you mirror each other, you'll be able to better understand each other and far more likely to like each other as well.[21]

We also suggest that you practice in front of mirror prior to giving an important talk. If you're going for a job interview, or need to present a new idea to boss, or to address a concern to a colleague, the time you spend rehearsing what you will say and coordinating your words with your body will help assure the best possible outcome. Therapists can become more effective, public speakers more highly rated, doctors more respected by their patients, managers more respected by their employees, and teachers better able to improve their student's work. As Spencer Kelly, a professor of psychology and neuroscience at Colgate University emphasizes, "Teachers can use gestures to become even more effective in several fundamental aspects of their profession, including communication, assessment of student knowledge, and the ability to instill a profound understanding of abstract concepts in traditionally difficult domains such as language and mathematics."[22]

Don't Be Tone Deaf When You Speak

Vocal inflection, like body language, also plays an essential part in conveying a message in a meaningful and persuasive way. As researchers at Emory University point out, the tone of your voice—the pitch, loudness, tempo, and rhythm—will often convey more useful information than the words you say.[23] Even dogs can discern the difference in your tone of voice, recognizing if your command is imperative, or simply informative.[24] And the same holds true for your wife and kids, and for the colleagues you interact with at work. If you don't use the right tone of voice, you may convey the wrong meaning, and thereby be responded to in a way that you did not intend.[25]

Vocal modulations convey emotional context, and they are so powerful that they can actually change the way that words and meanings are embedded into memory.[26] This gives us an important clue concerning effective communication. We want to make sure that people will remember what we have said, and vice versa, which means that we have to train ourselves to play close attention to all the elements of communication: words, tone, facial expressions, gestures, and other subtle cues.

That's a lot of information to take in, and the best way to do it is to slow down the conversational process, speaking for briefer periods of time and listening more intently to the other person's vocal inflections. However, if you are stressed out or in a hurry, you will likely ignore these suggestions, which is why we emphasize the importance of staying relaxed during every stage of the conversation.

What Do Words Taste Like?

Words, you may be surprised to hear, have different flavors, and speech can trigger sensations in our mouth and gut![27] High tones tend to taste sweet or sour, and low tones taste more salty.[28] As one Oxford scientist put it, bitter flavors taste like a trombone.[29] Studies like these show that communication is a multisensory process, and if we speak too fast, without awareness of our physical and emotional state, we may overlook cues that provide important information for solving problems and working with others.

The Brain's Compassionate Circuits

We're just beginning to map the social-communication circuits in the brain, but two of the newest evolutionary structures—the insula and the anterior cingulate—appear to work together when we engage in important social interactions. They are involved in the expression of compassion and empathy, in conflict resolution, and in the recognition of deception. They work in tandem to regulate our emotional reactions and behavior, they both play a major role in suppressing fear and anger generated by the amygdala, and they are also directly involved in the language processing, speech, and listening.[30]

These structures are essential for developing the skills of self-reflection and introspection, and they have strong neural connections to other major structures in the brain.[31] Many of the twelve strategies of Compassionate Communication have been shown to improve the functioning of these areas, thus increasing our ability to respond to other people with deep empathy and concern.[32] In fact, the types of mental practices included in this book have been shown to increase the size, thickness, and activity in both the insula[33] and the anterior cingulate.[34] This suggests that the improvements in communication may become permanent if you practice Compassionate Communication regularly.

Thinking in Pictures

Before we learn to think in words, we instinctively think in pictures. As the brain continues to develop, we gain the ability to think in increasingly abstract ways. The following illustration shows how we mature from the language of pictures to the language of words:

Picture Drawing Symbol Word

In general, pictorial language is processed in the rear regions of the brain, while abstract concepts engage the language regions in the frontal lobe.[35] To communicate effectively and have meaningful dialogue with others, we need to use a combination of words, symbols, and images.[36]

When Things Go Wrong in Our Brain

The neuroscience of communication is one of the most complicated fields of research because there are so many elements involved. Different parts of the brain are constantly "talking" to each other and "relating" with each other, but if one small part is damaged, our capacity to communicate effectively with others can collapse.

Let me give you several personal examples I, Andy, encountered when I was a resident in the neurology wing of a university medical center. One of the first patients I was assigned to was John. He'd had a stroke that affected the part of his brain associated with "receptive language." When I walked into the room, he immediately struck up a conversation. He told me how well he was doing and how quickly he was recovering. He even complimented me because of the excellent care he was getting from the hospital.

"But," he added, "the food is lousy."

I was impressed. After all, chatty patients are often the first ones to recover from their neurological problems.

"Are you having any pain?" I asked.

"Well, the nurses are nice and really attractive!"

This was not the answer I expected. So I asked him another question. "Do you know what hospital you are in?"

"You see that other patient in my room? He snored throughout the night and kept me awake."

No matter what question I asked, his response was unrelated. What was happening? The areas involved in listening had been damaged, but the areas involved in speaking were just fine. He could tell me exactly what he was feeling, and he could talk to me about anything *he* wanted to talk about, but he had no ability to process what I was saying to him. Even if I gave him a simple command like "raise your arm," he wasn't able to respond. He wasn't angry or frustrated with me, but I felt like I was talking to a brick wall.

The next patient I examined was Sally, a seventy-four-year-old woman who'd had a stroke near Broca's area, the part of the brain that allows us to construct sentences and make proper use of them. I went in, introduced myself, and asked her how she was feeling. She just nodded. But when I asked her to raise her left hand, she did it right away.

"Tell me your name," I asked.

Again all she did was nod.

"What day is it?"

Another nod.

Then I asked her to clap her hands, and she immediately put her hands together and started to clap.

This patient was unable to express herself in words. She could understand what I was saying, and carry out any action I requested, but she couldn't access her language centers to express what she was thinking. This condition is called expressive aphasia.

On the next day, I visited Michael, a sixty-eight-year-old gentleman who also had expressive aphasia. He didn't have any of the problems Sally had. Instead he called things by the wrong name.

I showed him my necktie and asked him what it was.

"It's a machine gun," he said.

Now that was surprising! I wanted to know whether he really thought it was a weapon, so I took it off and handed it to him.

"Tell me how you use it," I asked.

He immediately took the tie, put it around his neck, and made a perfect Windsor knot. Clearly, he knew what it was, but his brain did not allow him to say the word "tie."

I asked him why he had called it a machine gun, and he gave me the strangest look. "I couldn't have said that! I must have been joking around with you."

Obviously, he had some awareness of his problem, and so he used the logic centers in his brain to cover up his embarrassment.

Some people have damage to the area between the listening and speaking areas of the brain. One of my patients, named Dorothy, had this problem after suffering a stroke. If I asked her to get a glass of water, she'd go to the sink, pick up a glass, and fill it. But if I asked her *what* she wanted to drink, she couldn't respond. Instead she'd say something like, "I don't feel like taking a walk."

Cases like these demonstrate how complex and fragile our language systems are. In certain forms of schizophrenia, the processing centers can become so scrambled that words come out in the most bizarre ways. And yet they seem to make perfect sense in the patient's mind.

Every human brain, from the moment of birth, develops in unique ways, and thus no two people have the same communication style. This allows for a wide range of creativity, but it also explains why it's easy to misunderstand one another. To make matters worse, we can only be consciously aware of a small amount of the inner communication that is constantly taking place. Research clearly shows that we all have the capacity to improve our awareness and our styles of communication, but we have to work at it every time we enter a dialogue. Consciousness demands that we stay in the present moment and not be distracted by the inner commotion of the brain and the outer commotion of the world.

To use the United Nations metaphor with which we opened this chapter, consciousness is like a single person assigned to translate a thousand different languages that are simultaneously being spoken by a thousand different speakers reporting on a thousand different subjects—all of which have an immediate bearing on your life.

If a government bureaucracy were run this way, the nation would probably collapse. But the human brain seems to do a fairly good job. Still, there's plenty of evidence showing that the limitations of human consciousness cause us undue stress and anxiety. Although this is the dilemma we face, we can train our brain to process information more effectively. We can move from the language of everyday consciousness to the language of transformational awareness, which is the topic of our next chapter.

CHAPTER 4

The Language of Consciousness

Awareness. Attentiveness. Alertness. Wakefulness. Intelligence. Self-reflection. Mental representation. Self-recognition. Symbolic association. Active thinking. Learned behavior. Linguistic understanding. Cognizance. Experience. Imagination. Internal testimony. Comprehension. Introspection. Personal identity. Remembering. Predicting. Imitation. Mind. Free will. Moral conscience. Inner speech. Explicit memory. Temporality. Subjective imagination. Analogy formation. Intentionality. Endogenous feedback. Rational control. Self-arousal. Mental time travel. Emergent creativity. Qualia. Universal being. God.

There you have it: a partial but concise summary of twenty-six hundred years of philosophical, theological, psychological, and scientific speculation on the nature of human consciousness. Everyone agrees that it exists, but so far no one knows what it is, where it is, or how it works.[1]

We don't even have an agreed-upon definition, which prompted William James—the father of American psychology—to conclude, in

1904, that consciousness was nothing more than "a mere echo, the faint rumor left behind by the disappearing 'soul' upon the air of philosophy."

A hundred years later, Nobel laureate Francis Crick and neuroscientist Christof Koch expressed a similar sentiment when they implored scientists to stop using the term, "except in a very loose way."[2] Yet they, like so many others, continued to write dozens of articles and books in an attempt to delineate the nature of this mysterious beast. Today the search for a definition continues, and it continues to be one of the hottest topics in science, psychology, and theology. Even the Vatican has chimed in, suggesting that consciousness is divinely bestowed upon us at the moment of conception. In this new field that some call "neurotheology," consciousness is the essence of one's soul.

Depending on how you define it, consciousness may not be unique to humans and may even be found in the most primitive single-celled organisms. For example, you can train bees to recognize color-coded symbols that direct them to turn left or right as they fly. They, like many insects, have long-term and short-term memories, qualities that are essential in human consciousness. Bees grasp abstract relationships, make group decisions, and have communication skills that rival human beings. In fact, neuron for neuron they hold more information than we do.[3]

We can't definitively argue, like we once did, that human consciousness is superior. Dolphins and whales, for example, exhibit language and social skills that surpass those of human beings in many ways, and various primates also appear to have more complex versions of consciousness.

Consciousness is neurologically entwined with the workings of nearly every part of the human brain,[4] and it allows us to be socially aware and communicative with others.[5] But there's always been the chicken-and-the-egg question: is consciousness a by-product of brain

Conscious Slime?

A slime mold is a funguslike organism that is neither plant nor animal. And yet this single-celled creature has enough intelligence to maneuver its way through a laboratory maze by mathematically calculating the shortest route to take. Should this not be considered a primitive form of consciousness?

activity, or does consciousness shape the brain? We now know that both are true. A single conscious thought can initiate activity throughout the entire brain,[6] and as our neuroimaging studies have shown, even the act of contemplating the nature of consciousness—as a group of Buddhist meditators did in our lab—is enough to alter both the structure and functioning of the brain.

Every year dozens of new studies appear that expand our understanding of human consciousness. Yet its nature and source remain a mystery. Because of this continuing elusiveness, scholars like Roger Penrose and Stuart Hameroff have suggested that the principles of quantum mechanics may be the best way to explain it.[7] So far no definite evidence has been found to support this intriguing idea, but it wouldn't surprise us if a connection was one day found.

However, there is substantial evidence to suggest that human consciousness—or something that appears to be related to intentional thought—can transcend what we normally assume to be the physical limitations of the body and the mind.[8] It may not be enough to have any practical use, but when Dean Radin, senior scientist at the Institute of Noetic Sciences, did a double-blind study on the effects of human intention on another person's autonomic nervous system, his team discovered that the sender's compassionate thoughts generated small changes in the skin conductance of the distant receiver, even though the person receiving the thoughts was unaware of the experiment.[9]

Radin's team also showed that our thoughts might be able to affect inert substances at a distance—such as water crystals that were located, literally, on the other side of the planet.[10] It's too early to evaluate the validity and consistency of such studies, but their findings strongly suggest that the brain's ability to communicate extends far beyond the boundaries of normal human interaction, defying any established scientific principle.

Consciousness and the Brain

If we want to understand the power of language and human communication, we have to include what we currently know about the nature of conscious thought. Consciousness, as far as the most recent research shows, begins the moment we come out of the womb.[11] Prior to birth the fetus is almost continuously asleep, with very little neural activity occurring in the areas that produce language.

But newborns immediately become aware that they are separate from other people and objects in the world. In other words, they have a sense of self and other. They also have a primal awareness that they need to communicate to others if they are going to survive in the world, and they do so through their vocal cords and body language. They scream, they smile, they wave their arms around to communicate their basic needs, and they exhibit emotional responsiveness and signs of shared feelings. They remember sounds and vowels that they heard in the womb,[12] and they exhibit spontaneous neural activity that corresponds to what William James called "the stream of consciousness."

Many of the structures that govern conscious speech reside in the outermost layers of the brain, but these areas are largely undeveloped at birth. Rapid neural growth begins immediately after we are born, as dense neural connections are made between the neocortex, the thalamus, and other deep structures of the brain. These changes predict the degree of consciousness that infants and children have, and consciousness continues to develop and change throughout one's life span.[13] Disrupt any part of this delicate circuitry, and conscious awareness can be permanently impaired.[14]

How Thoughts Become Real

In the center of our brain there's a walnut-shaped structure called the thalamus. It relays sensory information about the outside world to the other parts of the brain. When we imagine something, this information is also sent to the thalamus. Our research suggests that the thalamus treats these thoughts and fantasies in the same way it processes sounds, smells, tastes, images, and touch. And it doesn't distinguish between inner and outer realities. Thus, if you think you are safe, the rest of your brain assumes that you are safe. But if you ruminate on imaginary fears or self-doubt, your brain presumes that there may be a real threat in the outside world. Our language-based thoughts shape our consciousness, and consciousness shapes the reality we perceive. So choose your words wisely because they become as real as the ground on which you stand.

Consciousness is a world unto itself—an abstract mental representation of an outside reality that we can never fully grasp. Take color, for example. It doesn't actually exist in the world. Light waves exist, but they are not what we "see" inside our brains. The brain's visual centers interpret the effects of the light waves on the color cones in our eyes, and the information is then reconstructed into an internal palette of colors, and it is categorized by the way we use our language.[15] Because human beings share the same visual functions, we all see a blue sky on a sunny day, even though the sky is not really "blue." But if you don't give that visual experience a specific name, the brain might not be able to "see" that specific color.

Color is also influenced by the culture in which we are raised, and if you grow up in a different country—be it Russia, England, or Africa—the words you assign to colors will alter what you actually see.[16] For example, members of the Berinmo tribe of Papua New Guinea cannot distinguish between blue and green. But they can be taught to do so, demonstrating that the perception and categorization of color is a language-bound category controlled by cognitive processes unique to human brains.[17]

The same is true for words. Change the sound or the intonation and the entire meaning can differ. When we talk to others, we need to keep this in mind, because different people can react to the same word or phrase in different ways based on their culture or childhood experiences. For example, a statement like "you are beautiful" can be viewed as a compliment by some people or as an invasion of privacy, especially by someone who may have experienced sexual abuse. In China, to tell someone that they are beautiful would be considered impolite.

Normally, when we speak we make the erroneous assumption that other people relate to our words in the same way we do. They don't. Thus we have to expand our consciousness about language to include the fact that everyone hears something different, even when we are using the same words. Words are needed to create our own inner reality and map of the world, but everyone creates a different map. To put it another way, consciousness—and the language we use to convey our feelings, thoughts, and beliefs—is a very personal and unique experience.[18] When we recognize this neurological fact, we become better communicators because we don't assume that other people understand what we say.

The Limitations of Everyday Consciousness

Scientists have identified many different levels, or states, of consciousness, and each one is governed by distinct neural networks in the brain.[19] But the one that concerns us the most is "everyday consciousness," which is different from other forms of self-reflective awareness.[20] Everyday consciousness consists of all the ordinary thoughts, feelings, and sensations we are aware of in any given moment, and it represents a very limited view of reality. It's like a snapshot, a tiny picture taken of an enormous panoramic view, and the information it contains changes from one moment to the next, altering our perception of the world.[21]

Everyday consciousness relies strongly on short-term working memory. We use it to form meaningful sentences and convey them to other people. But as we mentioned in the opening chapter, the average listener can only pay attention to a small amount of information for a brief period of time. When we consciously want to communicate something to someone else, our working memory selects about three or four "chunks" of data at a time, pulling it from our vast reservoirs of stored information.[22]

What's a chunk? It's a small packet of related information that the brain has chosen. It represents a specific feeling, idea, or thought, and short-term memory can only hold on to those chunks of data, on average, for about twenty to thirty seconds.[23] Then they get dumped from working memory as new chunks of information get uploaded. It's like looking at a vast landscape of trees, rocks, vegetation, chirping birds, and sunlight filtering down through the leaves. We can't consciously pay attention to all the details, so the brain consolidates the information and calls it a "forest." It picks one word to represent a complex experience, uses it to deal with the immediate situation, and then forgets about it as it loads the next four chunks of information into working memory.

And that's exactly what happens when we listen to another person speak. The brain takes all the words and implied meanings and summarizes them into a momentary thought. If the person who is speaking uses too much information, our unconscious processes arbitrarily choose which words seem relevant in that moment.

You can guess what the problem is. Most of us believe it's best to give a detailed description to the person with whom we're conversing, never realizing that they can only focus on four tiny chunks of information and for a very brief period of time.

Let's use the previous paragraph as an example. You can say it in about ten seconds, and you can read it even faster. But I bet you wouldn't be able to repeat it, even if you reread it a dozen times. Why? Because it contains between ten and fifteen chunks of information, which is far more data than everyday consciousness can handle. When researchers at the University of Missouri tested young and old adults, they found that even a single sentence composed of ten words was difficult to recall accurately.[24]

When we understand the limitations of everyday consciousness, we can use this information to become better communicators by speaking briefly and then asking the person if they understood what we said. If the concept you want to convey is new or complicated, then repeating your message in different ways will help the other person's brain to build an inner comprehension of its essential elements.

We can also improve our communication skills by taking advantage another neuroscientific fact: the slower we speak, the more the listener's comprehension will increase.[25] Speaking slowly also relaxes both the speaker's and listener's bodies.[26] The result? Less stress and greater understanding, with the least expenditure of words. It's a win-win situation—for your body, your brain, and each other—and the formula is easy to remember:

KEEP YOUR SENTENCES SHORT
AND SPEAK SLOWLY, FOR THIRTY SECONDS OR LESS

Brevity Reduces Conflict

Sometimes even thirty seconds is too long, especially in situations where emotions run high. When Mark introduced Compassionate Communication at a meeting of the Coalition for Collaborative Divorce (a Southern California organization of collaborative attorneys, therapists, and financial advisors that helps to mediate divorces peacefully, thus keeping the couples out of the courts), he created a role-playing scenario. Two individuals acted out a hostile confrontation in which neither party was willing to compromise on a settlement issue. Different attorneys tried different strategies but to no avail, because both parties continued to argue and defend their positions. And even though they were role-playing, you could feel real tensions rising in the room.

Even limiting the dialogue to thirty seconds did not help, so we changed the rule. Everyone, including the attorney in the role play, was limited to one sentence, lasting ten seconds or less. Within five minutes, the attorney was able to move the stalemate to a point of mutual agreement. By severely limiting communication, the ability to express anger was removed. That's the funny thing about anger. Not only does it give the person a false sense of self-righteousness, it neurologically generates more anger.

When we work within severe time limits, sometimes the most creative and utilitarian ideas pop up. When we suspend the mental chatter of everyday consciousness, other forms of cognitive processing come to the foreground. These deeper intuitive capacities involve different neural processes, and they assess a situation faster than the mechanisms that make use of the "chunking" mechanism of working memory.

Stay Relevant and Avoid Distracting Sounds

When we consciously limit our speaking to thirty seconds or less, we learn to select our words more carefully. This approach has a specific neurological advantage, because irrelevant speech disrupts patterns of neural coherence in the brain, making it difficult for the listener to understand what is really being said.[27]

Irrelevant speech also interferes with judgment and learning.[28] Scientists at the Air Force Research Laboratory in Ohio found that when two people talk at the same time, it degrades a person's ability to pick up important verbal cues.[29] In fact, any background conversation—as when you're sitting in a restaurant or a cubicle in a crowded office—will interfere with your brain's ability to carry out any form of mental task.[30] Even hearing traffic noise in the background is enough to impair a person's ability to learn.[31] Our advice: if the conversation is important, find the quietest place possible so that you can fully concentrate on each and every word you hear.

Becoming Conscious of Everyday Consciousness

Compassionate Communication teaches us how to become acutely aware of the way we normally think, and the moment we turn our attention toward the mind's inner processes, the molecular, cellular, and chemical functioning of the brain begins to change.[32] In essence we create a new form of consciousness that is based on self-reflection and observation.

When we turn our awareness onto ourselves, we often make an astonishing discovery: it's nearly impossible to stop our minds from thinking. In fact, everyday consciousness appears to involve a continual stream of inner dialogue, and we can tune into that dialogue in the same way we listen to other people. There's even a recognized term for this ongoing neurological process. It's called "inner speech," and it can pose a real problem by interfering with our ability to pay attention to what other people are saying.

The Inner Voice of Consciousness

Inner speech preoccupies most of our waking life. It gives voice to our *interior* experience of the world around us,[33] and as researchers at the University of Toronto have found, "the inner voice helps us to exert self-control by enhancing our ability to restrain our impulses."[34] In fact, higher frequencies of inner speech are associated with lower levels of psychological distress.[35]

In 1926 the famous Swiss psychologist and educator Jean Piaget noticed that many children begin to talk to themselves between the ages of three and five. For example, when a child builds a house of blocks, she might often verbalize her actions: "Now I'm going to put the red block on top of the blue block." When she's done, she might say, "Now everything will fall down," as she pushes the blocks over.

Piaget called this activity "egocentric speech," and it demonstrates how language begins to dominate our daily lives. We use inner speech to make conscious decisions and to shape our thoughts in ways that help us communicate them to others, and we use inner speech to rehearse what we are about to say.

Inner speech begins in the first few years of life, and we continue to have these internal dialogues throughout our life span.[36] It appears to occur in the left hemisphere of the brain—the side where abstract language is processed—and it plays a specific role in orienting us toward other people in the world.[37] Inner speech also helps us to regulate our awareness of ourselves.[38]

When you pay close attention to your inner speech, you'll discover that each emotional state—anger, fear, depression, joy, contentment, etc.—has its own voice and style of communicating. If you think this sounds like having a multiple personality disorder, you're not far from the truth, because we all have dozens of subpersonalities, and each one has a temperament of its own. Normally these inner voices blur into each other and are somewhat indistinct, but severe traumas can unglue these personalities, freeing them to act autonomously.

Although we may not always be aware of it, different inner voices are continually commenting on our behavior. Being self-critical is a perfect example: one part of us does a job and another part chimes in and whispers, "It's not good enough. The boss is going to complain." Nor is it unusual to find our different personalities debating. We see a piece of clothing in a store and fall in love with it, but then the voices start: "You can't afford it!" "But I deserve it!" and so on.

Each of these inner voices has a different effect on your brain. A self-critical voice will stimulate error-detection circuits, whereas a self-reassuring voice will stimulate the neural circuits involved with compassion and empathy.[39]

Negative inner dialogue can be particularly destructive. For example, anorexics often experience harsh and forceful inner voices.[40] It helps them to stifle their impulse to eat, but the negative inner speech continues to erode their self-confidence, so they end up starving themselves in life-threatening ways. When they learn to stop listening to these destructive voices, their eating behavior improves.

Workaholics suffer a similar fate. No matter how much they accomplish, the inner perfectionist won't let them rest: "You need to work more! This is not enough! What will happen if you fail?" To stop this type A behavior, which is damaging to both heart and brain, the obsessive-compulsive worker has to develop a new inner language that puts a high value on nonmaterial goals like friendship and on pleasurable pursuits.

Procrastination is another form of destructive inner speech: "What if I fail? I don't know enough to succeed. Oh heck, I can deal with these problems tomorrow." How do you stop this kind of inner speech? By deliberately interrupting it and replacing it with repetitive statements that bolster confidence and self esteem.[41] If you change your inner speech, you change your behavior, and you actually improve the functioning of your brain. The ability to control the tone of our inner dialogues is the first step toward winning the trust and respect of ourselves and others.

Observing Your Inner Speech

We all have the power to change our inner speech in ways that will improve our lives. But we must first learn how to listen with our "inner ear." This involves a different brain system from the one that controls inner speech.[42]

Here's a little exercise you can do right now that will help you to identify these inner voices and to distinguish the useful ones from the disruptive ones. Get a sheet of paper and a pencil, and find a quiet place to sit. Take a few deep breaths, then yawn and stretch for about twenty or thirty seconds. The more relaxed you are, the easier it will be for you to hear your inner speech.

Now just sit back in your chair and remain silent. Try not to think about anything. You'll soon discover that it's like the old parlor game where someone tells you to not think about elephants. Immediately an image of an elephant pops into your mind. Now get rid of that elephant and take another deep breath. Close your eyes and remain silent for as long as you can.

Most people quickly become aware of fragmentary thoughts drifting in and out of consciousness. When you notice these thoughts, write them down on the sheet of paper, along with any feeling or sensation you are aware of. Then let the thought float away, as if it were a cloud in the sky. This helps you to stay neutral as you observe the constant shifts in your awareness.

After you write down each thought, take a deep breath and relax, paying attention to whatever happens next. Continue to observe, note, and let go. The longer you do this, the more intense the experience becomes. Sometimes the periods of silence will increase, and other times they will decrease, to the point where it can make you feel like you're going nuts. As we said earlier, the brain doesn't like to change, and the inner voices certainly don't like being ignored!

But if you sit there and observe the voices without judging them—which is the most important part of this equation—you'll be developing a powerful psychological tool. As eight randomized controlled studies have shown, this exercise is one of the fastest techniques for reducing stress, anxiety, irritability, and depression.[43]

When you learn how to master deep awareness of yourself and others, you are less likely to get caught up in destructive emotional states that can sabotage your ability to communicate effectively and compassionately with others.

Transforming Negative Inner Speech

Inner speech is not necessarily bad. It helps us to manage strong emotional reactions, and it gives us the power to modify inappropriate behavior.[44] If you are feeling anxious, worried, or highly stressed, positive inner speech can help you to feel calmer.[45] And positive self-talk improves the performance of people engaged in active sports.[46] It's like having an inner coach, but you have to develop it and weed out the negative advice.

Let's say you're up to bat at the company-sponsored baseball game. Positive inner speech can be as simple as telling yourself, "I can do it!" or it can be as complex as devising a strategy to fake out the pitcher. But let's say you strike out. There's a tendency for an inner voice to blame you, or someone, for your defeat. This is the speech you need to interrupt, replacing it with assurance that you can do better the next time you go up to bat.

When inner speech turns negative—and it can happen to even the most successful people in the world[47]—it will, over time, generate a plethora of problems. It can stimulate eating disorders, passivity, insomnia, agoraphobia, compulsive gambling, sexual dysfunction, low self-esteem, and depression. It can make you quit your job in a self-destructive way, and it can drive you to treating your family with disdain.

On the other hand, positive self-talk improves attentiveness, autonomy, confidence, and work performance.[48] It doesn't seem the matter what the words are, as long as they are positive, repetitive, and realistic. And you have to use your words to generate a plan. For example, just wishing you'd make a million dollars won't make you a penny, but if you use positive inner speech to plot out a sound financial plan, you'll increase your changes of success enormously. The moment self-doubt creeps in, it will sabotage your drive toward achieving your goals and dreams.

Sara White, a professor at the University of California, San Francisco, and a distinguished leadership and communication coach for the medical community, recommends these steps for turning negative inner speech into positive self-talk. Doing so will help enhance your performance, satisfaction, and professional success.[49]

- Observe your inner speech and keep a "thought record"
- Confront your inner critic and rewrite self-limiting scripts
- Replace negative thoughts with positive inner dialogue

- Look for the gift and opportunity in every obstacle you meet
- Focus on your accomplishments, not your setbacks
- Review, reinforce, and practice your new self-talk

Obviously, self-talk serves many purposes, but it does have a hitch. It can distract us from paying full attention to what another person is saying. Instead of listening, we're often unconsciously rehearsing what we want to say next, and whenever our attention is split like this, it interferes with the processes that govern memory, cognition, and social awareness.[50]

Inner speech takes us out of the present moment. It may to be essential for solving difficult problems, but it can distract us from truly listening to other people.

The Voice of Insight

With enough practice, deep awareness of our inner speech can lead to a surprising discovery. A new voice will occasionally appear: the voice of intuition. Often it is preceded by a long period of inner silence. Then a sudden insight may burst into consciousness. In that moment you might get a glimpse of the larger picture.

This kind of intuitional insight is often difficult to put into words, but it *feels* true. It's the "eureka!" or "aha!" experience that has for eons been associated with creative flashes of insight. This too turns out to be a unique language-driven experience, one that is governed by the right hemisphere of the brain.[51] The insula and anterior cingulate, which also govern our feelings of compassion, are stimulated,[52] and in these moments of deep awareness gamma-wave oscillations in the brain spike as everyday consciousness falls away.[53] Cognitive restructuring takes place, and suddenly the world looks different.[54]

The feeling may last a second or stay with you throughout life, but anecdotal stories about the experience are surprisingly the same: a new consciousness emerges that allows one to function more fully, more efficiently, and with a deeper sense of personal satisfaction and inner peace.

The Language of Silence: Is It Really Golden?

So far we've given a lot of attention to inner speech and thoughts and the effects they have on consciousness and the brain. But what about silence? Does it have any neurological value? Yes. In fact, if you don't pause for a few seconds between each brief phrase, the listener's comprehension will decrease.[55] The same holds true when you use unfamiliar or technical words; it takes more time for the listener's brain to process them. Thus you need to speak briefly and then pause, leaving a few seconds of silence at the end of each phrase or sentence.

Great teachers, therapists, actors, and public speakers know the power of a silent pause, and they consciously use it in their work. Salespeople and CEOs do the same, because they know that such pauses create a deeper connection between people. They also know how important it is to let the other person talk: it's the only way to gain true insight into their desires and needs. Deep listening requires silence, which means we have to train our mind to distance itself from the inner speech it continuously generates.

This raises an interesting question: is it really possible to achieve a true state of inner silence? Yes, but only for brief periods of time. Even if you are placed in a soundproof space, the auditory part of your brain will immediately become activated, and you'll automatically hear verbal "messages."[56] In other words, the brain is not used to silence because from an evolutionary perspective too much silence can be a sign of danger. When the normal bird and animal sounds in a forest stop, it usually means there's a predator afoot.

Listening to the Silence between Words

Here's a little experiment we would like you to try. It's going to feel very strange at first, but it will help you to hear how busy your mind can be as it thinks up the words it wants you to say. You can do this exercise alone, but it's far more interesting to find a partner to experiment with.

All you have to do is to say a single sentence aloud. But we want you to pause for one second between each word. Then we want you to say another sentence, but this time leave two seconds between each word. With each additional sentence, pause for an additional second and notice how your inner speech begins to react. I assure you that it will become quite noisy and agitated.

Interestingly, when we conduct this experiment in workshops, the speakers usually feels increased anxiety, but the listener tends to feel increasing calm. To get a brief sense of this experience right now, read the following sentence aloud, pausing for four or five seconds between each word:

AS YOU SPEAK THIS SLOWLY,
NOTICE WHAT YOU EXPERIENCE
BETWEEN EACH OF THESE WORDS.

Read the sentence again out loud, with even longer pauses, and when you come to multisyllable words, say them even slower than you did before. Stretch out each word and pay attention to the sounds of the consonants and vowels. Instead of worrying about what the message is expressing, just take some time to notice the experience of speaking slowly.

Most people, when they do this exercise, will hear all kinds of inner commentaries in the silences between the words: "This is silly." "This is weird." "It's ridiculous to speak this way." Sometimes the inner speech slows down, but often it will speed up. It's as if the mind feels it has to talk fast so that it can fit in everything it thinks it is important to say. People often speak rapidly, and for long periods of time, for fear that if they don't describe everything in full detail, the other person won't be able to understand. But if you remember the golden rule of short-term working memory, you'll know that the other person is only going to remember a very small portion of what you say.

When you speak super slowly, you'll begin to use the silences to carefully select the next word. You can actually think about what you want to say while you're saying it. In a matter of minutes, you may even begin to notice that you can communicate a great deal of information with half the words you would normally use.

When doing this experiment with a partner, say only one sentence; then let the other person speak. Continue this slow-paced exchange for at least ten minutes. You'll find that listening to a person who is speaking slowly is rather enjoyable, and you'll soon have the sense that you are beginning to understand them with greater accuracy and depth. It's an incredible experience, so we strongly suggest that you try this exercise with several friends and family members, and then try it with several colleagues at work.

Here's what happened when Mark first tried this exercise with his wife, several years ago. The first few minutes were spent just getting used to the unusual rate of speech. Then Mark, speaking super slowly, asked Susan a question: "How . . . does . . . this . . . way . . . of . . . speaking . . . feel . . . to . . . you?"

"I . . . like . . . it," Susan responded just as slowly.

"Why?" asked Mark, after pausing for about five seconds.

"It . . . doesn't . . . make . . . me . . . nervous."

Mark slowly responded, "But . . . there's . . . no . . . emotion . . . in . . . my . . . voice."

I . . . know," said Susan. "When . . . you're . . . emotional . . . , I . . . sometimes . . . think . . . you're mad."

A long pause followed before Mark spoke. "My . . . mind . . . is . . . racing . . . right . . . now." Another long pause. "Without . . . emotion . . . , my . . . speaking . . . career . . . would . . . end."

Susan didn't respond.

Mark listened to his inner thoughts, gradually deciding which would be the most relevant to share. Trusting his intuition, he finally said, "Do . . . you . . . really . . . want . . . me . . . to . . . speak . . . this slowly?"

"Yes!" said Susan said without pausing.

"Why?"

"I . . . don't . . . know . . . I'm . . . just . . . noticing . . . it . . . for . . . the . . . first . . . time," she replied. "When . . . you . . . speak . . . normally . . . , I . . . get . . . anxious . . . by . . . the . . . amount . . . of . . . emotion . . . in . . . your . . . voice.

Mark thought a long time about that. It didn't make much sense, but, then again, if it distanced his wife from him, why not give it a try? "How . . . long . . . do . . . you . . . want . . . me . . . to . . . do . . . this . . . for?" he asked.

"Till . . . Christmas!"

"Five months?" exclaimed Mark, forgetting to pause between his words.

"Yes. . . , five . . . months," Susan replied with a smile.

All kinds of thoughts raced through Mark's mind: "Christmas? Five months? No way! Wow! I make my wife anxious? Wow! She never said that to me before. Her problem? My problem?" His mind continued to race, and then it eventually calmed down and turned quiet.

They agreed to continue talking super slowly, and although they didn't do it for more than one or two weeks, it was one of the most transformative periods in their relationship. That afternoon they talked for four hours about disturbing events that had occurred ten years earlier, and which they'd never been able to address with any mutual satisfaction. Over the next couple of weeks, they continued to resolve one conflict after the other, and they now have a formal agreement to speak very slowly to each other whenever a difficult conversation arises.

Emotions play an important role in the communication process, but when they are presented with too much drama, they can evoke defensive reactions in many listeners. Often we're not aware of the emotional impact that our speech has on others, and the example given above serves as a reminder that we should always get feedback from others. Ask them what you can do to become a better communicator, and remember that different people will feel comfortable with different styles of interaction. If we want to excel in our conversations with others, we need to realize that every dialogue is a unique experience, and that each dialogue—even with the same person—might require us to adjust our tone of voice and the time we give to speaking and listening to each other.

Furthermore, when conversing with people who may have deep inner wounds brought about from previous interactions, we may need to take even more time and care to ensure that our words do not push hidden buttons that the listener may be unaware of.

As an exercise, practicing slow speech is an excellent way to ensure that we are choosing the best words to convey what we really want to say. And when emotional buttons get pushed, if we consciously slow down our speech and use the warmest tone of voice possible, our words and body language will help the other person to relax. Trust and empathy can be undermined with a single negative expression, but it can be rebuilt if we consciously generate compassion for each other.

Improving the Silence

When you consciously learn how to be a silent observer of your own thoughts and feelings, a different type of silence emerges. In that silence people often have a curious insight. They become aware that they are watching themselves. But then another thought occurs: who is this person who is doing the observing? This "self," if we can call it that, is different from all the ideas we normally have about who we are, and it tends to be calm, serene, and mostly silent. It watches but doesn't react. It listens and rarely speaks, but when it does, most people experience it as a form of inner wisdom.

In the silence created by this unique form of awareness, we can improve our ability to make predictions about the future, and this allows us to make better decisions concerning our work and our life.[57] According to researchers at the Neurosciences Institute in San Diego, the observing self "appears to be needed to maintain the conscious state."[58]

It's an interesting paradox: we need to have an observing self to be conscious, but most of us are unconscious of the observing self! Instead we give far more attention to the more superficial self-image of who we think we are. These impressions are filled with our fantasies and judgments about who we want to be and who we fear we might be, but none of these ideas is accurate. When we learn how to use our observing self to watch these other images, we begin to realize that they aren't necessarily real. They're just opinions—from ourself and others—that we've come to accept over the years. The emerging research on consciousness suggests that the observing self can take a more accurate view of reality. It doesn't seem to get upset like our normal selves do, and the more we reflect on this deeper form of awareness, the less anxious and depressed we become.[59]

We're not born with a conscious ability to observe our own awareness, but we can develop this skill by using the exercises included throughout this book. A strong, observing self actually *predicts* enhanced well-being.[60] It lowers emotional stress,[61] and it makes us more socially aware of other people's needs.[62] That's why we consider deep self-reflection an essential component of Compassionate Communication.

In truth, we really don't have to talk as much as we think we do. Mostly we just repeat the inner speech our brain is using to consolidate the overwhelming amount of information that is flowing into consciousness. If we take a few moments to step back and observe this inner world, we'll discover that most of the words we hear with our inner ear do not need to be shared with others. They have their own inner dialogues to engage in.

But if we want to have more productive and meaningful outer dialogues, both the speaker and listener need to slow down enough to allow the inner wisdom of the observing self—one's intuition—to emerge in the brief periods of silence we create. In that improved state of consciousness, we'll choose our words more wisely.

As a shrewd Hasidic rabbi once said, "Before you speak, ask yourself this question: will your words improve the silence?"

Chapter 5

The Language of Cooperation

If we were totally selfish, isolated creatures, there would be little need for communication. We'd simply do what we wanted to do, whenever we wanted to do it. But if every living organism were to engage in such behavior, competition over scarce resources—like food, water, or viable mates—would immediately escalate into violent conflict. Throughout the natural world, biologists have identified thousands of interrelational strategies that are designed to keep the peace. Those strategies can be boiled down to two words: "cooperative communication."

In order to survive, there has to be a balance between how much we take, how much we share, and how much we give to others who cannot fend for themselves. But the question remains: are humans inclined to be more selfish or cooperative, more greedy or generous?

When we first began to develop this book, we were inclined to believe that humans were fundamentally selfish. In fact, one of the early working titles was *The Selfish Brain*, a paraphrase of the classic work by Richard Dawkins, *The Selfish Gene.* There's plenty of evidence to support the argument for selfishness, but years of research has convinced us that the opposite is true. Only as infants are we allowed the freedom to be utterly selfish. Our brains are so undeveloped at the time of birth that we must depend on caregivers to provide us with every need.

However, our selfishness doesn't last long, for as soon as we can care for ourselves, our family members demand reciprocation. We have to learn to share our toys with siblings and friends, we have to do chores for our parents, and we have to limit our selfish impulses when we enter school. If we don't, we're punished. We're sent to our room or to the corner of the classroom, deprived of social contact, and this painful message makes it clear that selfishness is rarely tolerated in the social arena of life.

Still, there remains an inner struggle. If we have to share something we value, a plethora of questions arises. For example, how much do we need to share, and for how long? This brings up other questions concerning degrees of fairness and generosity, but there are never clear answers to guide us. Since each situation is different, involving different people with different notions concerning these values, we have to turn to our words and negotiate agreements. If we fail to find a mutually satisfying solution, the other person will not cooperate with us. The same holds true for work. No one is going to hire us and give us money unless we can give them something of value in return.

There is no *language* of selfishness. When we're selfish, there's no exchange of property or words. We simply take what we want without asking. But fairness requires cooperation, and cooperation is entirely dependent on a combination of dialogue, bargaining, compromise, and behavioral change. These are the basic elements that have been considered by two new fields of research: neuroeconomics and social neuroscience.

By placing animals and people in brain-scan machines as they engage in a variety of monetary exchanges, we have discovered a fundamental fact about human nature: in social situations, we reward helpful people with kindness and generosity and punish the unhelpful ones, even when the punishment incurs some cost to ourselves.[1] And the more we see people behaving in ways that are fair, cooperative, and kind, the more willing we are to form long-term friendships with them.[2]

Do All Organisms Communicate and Cooperate?

Human beings aren't the only organisms on this planet to cooperatively communicate with others. As biologist Joel Sachs, at the University of California, Riverside, reports, cooperation "pervades all levels of biological organization."[3] Even the lowly bacterium exhibits astonishing social behavior that is governed by specific forms of chemical communication.[4] In fact, plants can communicate with each other in ways that are remarkably similar to humans. Poplar trees, tomatoes, and lima beans communicate among themselves—through the air and through their roots—and they can even communicate with different plant species, animals, and microorganisms.

Like humans, plants use their communication strategies to cooperate with each other and to protect themselves from enemies. For example, some plants can literally cry out for help when they're being eaten, and the signals they emit can attract carnivorous enemies of the animal grazing on them.[5] Some plants even appear to have the capacity to listen, while others appear to be deaf.[6] They don't use words, but they do have signaling receptors and pathways that are similar to the communication networks that occur in our brain.[7] And they even have their own form of inner speech. For example, some plants can use their vascular networks to send hormonal signals to other parts of the plant.[8]

Biologists call this process "intraplant communication," but for us, it's a reminder that communication takes place on many nonverbal levels, not just in plants but in human beings as well. In his book *Gaia*, James Lovelock even suggested that the entire earth represents a living organism, with its own system of communication that works to create an integrative, cooperative organism. However, unlike plants and most other living creatures on earth, only we, as humans, can consciously choose to increase our levels of cooperation by changing the *way* we communicate to others.

Human Cooperation and Neural Resonance

Boiled down to its essentials, communication involves the accurate transference of information from one brain to another. We do this through the process called neural resonance, and the more we can mirror the neural activity in other person's brain, the better we are able to cooperate with them. If we closely observe a person's face, their gestures, and their tone of voice, our brain will begin to align with theirs, and this allows us to know more fully what the other person is thinking, feeling, and believing.

Researchers at the Social Brain Laboratory in the Netherlands demonstrated this by having couples play the game charades. One partner was strapped into an fMRI scanner, and a word was shown on a screen. The person made hand gestures to describe the word, and a movie was made of the gesturing. Then the other partner went into the scanner, watched the movie, and tried to guess what the word was. The results? Similar areas were stimulated in both the sender's and receiver's brains when the latter guessed correctly,, especially the areas involved in language recognition and speech.[9]

This tells us several important things: that words can be conveyed through specific gestures, and that both the gesture and word stimulate similar areas of the brain—areas associated with language comprehension. In fact, as researchers at Max Planck Institute for Human Cognitive and Brain Sciences report, "Hand signs with symbolic meaning can often be utilized more successfully than words to communicate an intention."[10]

Most important, these studies reconfirm what we've been saying about neural resonance: if you really want to understand what the other person is saying, you have to listen to and observe the other person as deeply and fully as possible. Otherwise your brains won't mirror each other. If we can't simulate in our own brains what another person is thinking and feeling, we won't be able to cooperate with them.

To achieve optimal cooperation, it also helps to have belief systems in alignment. If they're not aligned, to the parties must first dialogue with each other to find common ground by agreeing on a mutually shared goal that will compensate both sides as fairly as possible. If common ground cannot be found, the communication process collapses, and any hope of cooperation will dissolve. As researchers at the University of Geneva have shown, when there is congruence of personal goals the brain systems relating to cooperation will be stimulated.[11]

The strategies of Compassionate Communication are designed to create neural resonance between two people as they converse, and if you enter a conversation with the *intention* of creating a fair exchange, you'll stimulate the cooperative circuits of your brain. The newest research also shows that the more you imitate, or mirror, a person's communication style, the more you'll increase the neural resonance between you and them, and this will generate more empathy, cooperation, and trust.[12]

Other research suggests that when you put yourself in a cooperative, empathetic state of mind, your emotional state may become socially contagious, spreading through your home, your work environment, and even your community like a virus.[13] In essence, we can bring people into alignment with our values and goals by spreading compassion nonverbally. Of course there are other ways to ensure cooperation, like using coercion, but the risks are higher because resentment quickly builds to the point where a person will be willing to make enormous sacrifices rather than submit to continuing abuse. We see this happening throughout the world today, as suppressed societies begin to demand equality, fairness, and justice from those who act with undue selfishness and greed. In these scenarios neural and cognitive *dissonance* has increased to the point where people can no longer remain passive.[14] Even small discrepancies between personal values and goals can undermine the communication process by creating chaos within the brain.[15]

Effective communication depends upon neural resonance. As researchers at Princeton University demonstrated in an fMRI brain-scan experiment, neural coupling vanishes when participants communicate poorly. In fact, they could even predict the degree of communication success by observing how closely one person's brain resonated to the other.[16] The researchers also discovered that good listeners—the ones who paid the closest attention to what was being said—could actually anticipate what the speaker was going to say a moment before they said it. Perhaps this is what a good psychic does: they pay attention to every nonverbal detail and use them to infer what the person is thinking and feeling. It's not magic; they're simply using some of the strategies of Compassionate Communication.

Mirroring Each Other's Voice

If you match the language style of the person you are conversing with, and the intensity of their voice, they will perceive you as being more related and attuned.[17] According to researchers at the University of Hawaii, "When partners interact, if things are to go well, their speech cycles must become mutually entrained." Not surprisingly, this verbal entrainment will improve your chances when you are searching for a suitable date. In an experiment involving speed dating—where you are only allowed four minutes to converse before you must switch to another person—couple with matching language styles were three times more likely to pair up by the end of the event. [18]

When two people like each other, they'll mirror each other's posture, facial gestures, and movements.[19] It's a sign that they feel connected,[20] it builds mutual rapport,[21] and it communicates a desire to affiliate and cooperate with each other. It may even earn you more money at work. When waitresses mirrored their customer's comments, they increased their tips by 50 percent.[22] Research has even shown that matching language style will improve the likelihood of a peaceful resolution, even in situations that involve serious conflicts and potential threats to one's life.[23]

Training Your Brain to Connect

If you want to increase your ability to resonate and empathize with someone else, just use your imagination. When a person is speaking, imagine you are them. Mentally visualize yourself in the situation they describe, and put in as much detail as possible, as if you were actually there. According to researchers at the University of Chicago, this form of mental simulation allows your brain to build a better understanding of the other person, and it doesn't matter if what you imagination is accurate.[24]

It even works with novels and movies, for the more you can project yourself into the role of the character, the more you'll feel compassion or, in the case of a villain, fear and disgust.[25] As António Damásio and his research team at the University of Southern California emphasize, when one actively projects oneself "into the shoes of another person, imagining someone's personal, emotional experience as if it were one's own," one triggers "the neural mechanism for true empathy."[26]

Can we train ourselves to feel compassion toward everyone? Yes, but it appears that we have a neurological mechanism that stops us from empathizing with people we don't like or respect. This "antimirror neuron" activity, as some researchers call it, appears to deactivate the brain's propensity to imitate another person.[27] Thus when we interact with someone whose behavior violates our personal ethics and beliefs, our empathy circuits shut down to ensure that we do not engage in similar acts.

There's even evidence suggesting that the more empathetic we are, the more accurate we become at predicting the other person's ability to engage in cooperative behavior.[28] But empathy has its limits. For example, we do not have the neural capacity to recognize when we have misread verbal and emotional cues.[29] Thus it is easy to think we understand what another person says and means, when in fact we don't.

Our advice: never presume that you know what a person really feels and means. The day I, Andy, got married, the rabbi kept repeating to us, "Never assume you understand what the other person is thinking—always make sure you ask and find out." The best way to do this is indeed to verify your assumptions with a question. For example, you might say something like "John, if I understand you correctly, I think you mean . . . Is that right?" If the other person doesn't agree, they will appreciate the opportunity you've given them to communicate what they really meant.

The Social Rules of Engagement: Anger Never Works

What happens when people don't cooperate, and how does the brain respond when somebody treats us unfairly or takes advantage of our generosity? We react with a well-documented biological process called "altruistic punishment." In fact, it turns out that the human brain is designed to initiate punishment whenever someone violates a social contract or behaves in a way we consider to be socially irresponsible.[30]

But there is a problem: violators don't appreciate being punished, and they are often unaware that they have violated the other person's trust. If you reprimand them, they'll feel resentful, the possibility of cooperation will deteriorate, and you'll run the risk of retaliation.[31] But if you don't say anything, the unfair behavior continues. In fact, if your voice shows even the slightest amount of disdain or sarcasm, it will be interpreted by the other person as an act of hostility. The result: relationship dissatisfaction and instability.[32]

In personal relationships, punishment—whether in the form of anger, criticism, or judgment—rarely works. But the brain seems to be hardwired when it comes to disappointment. If we don't get what we want—even if what we want is unrealistic—the brain's anger center gets stimulated. If our desires are frustrated, and the reward we hope for is postponed, the anger center gets stimulated. If we're in a rush and someone in front of us is driving slowly, we get irritated because our selfish desires are thwarted.

The best solution to the cycle that we know of is to interrupt the negativity by generating a thought that expresses compassion for yourself, the situation, and the other people involved. The research is robust: if we deliberately send a kind thought to the person we perceive as having violated our personal space, we psychologically increase our sense of social connectedness and strengthen the neurological circuits of empathy and cooperation.[33]

Researchers at the Program for Evolutionary Dynamics at Harvard University have found that people who use punishment the least are more likely to gain more cooperation from others, as well as to increase financial benefits for themselves. They state bluntly, "Winners don't punish . . . while losers punish and perish."[34]

Power Plays Don't Work

According to the United Nations, cooperation, not *power*, is the key to conflict resolution. When one party tries to impose its belief systems and values on another, conflicts escalate. If a dispute is settled through coercion, both parties feel less satisfied with the outcome.[35]

Research at Cornell University's Department of Neurobiology and Behavior found that there is something else you can do to improve your chances of forming stronger cooperative relationships with others, at home and at work: be more generous.[36] Generosity sends a specific message to other people's brains, telling them that you intend no harm. If a disagreement is being aired, it de-escalates the potential of an angry rebuttal and opens the door to reengage in a cooperative conversation.

In other words, being kind to those who are unkind to you will soften their hearts and soothe their angry brains. So the next time someone zooms up behind you, blasting their horn and waving their frustrated hands at you, give them the right of way. By pulling over and letting them speed by, you have given shown a little bit of respect. And perhaps one day they will return the favor to someone else.

The same thing applies to conflicts at work. If you show your unkind boss a little extra compassion, your financial security will remain intact. Kindness builds cooperation, and cooperation builds a better brain.

Chapter 6

The Language of Trust

The brain is an enormous communication center. It has approximately eighty-five billion neurons and eighty-five glial cells, each with a "mind" of its own. They're efficient and cooperative and continuously connecting and disconnecting from one another.

This is the storage bin of our accumulated knowledge, feelings, memories, beliefs, and habituated behaviors, but only a tiny bit of this vast information is accessible to our everyday consciousness—those four or five chunks of information that we can hold in our working memory for a half minute or less. Compared to the rest of our brain, conscious awareness seems highly inefficient. It has a very limited view of reality, and it compensates by filling in with a lot of assumptions and guesses.

How accurate is this model of the world consciousness constructs? We really don't know, and so we basically have to trust that it is reliable and useful. For the most part, it does a reasonably good job, but even so, it offers us only very limited access to the reality going on in other people's minds. When we engage in a meaningful conversation, our brains attempt to evaluate the trustworthiness of other people's intentions and words. If we can't establish trust, we can't do business, and we certainly won't fall in love.

First let's briefly define "trust." The dictionary gives us a range of choices: hope, faith, belief, reliance, confidence, and dependence. In relationships trust is the confidence we place in another person who we believe we can rely on in order to achieve a cooperative goal. But trust is something that cannot be measured directly. We can measure money, and to a large extent performance, but what quantity of trust does it take to assure that a successful exchange will occur?

Since our brain does not place much trust in the words other people use, it looks for other ways to gauge trustworthiness. We try to determine a person's character by evaluating their performance, abilities, and strengths, but the brain also gives special attention to their eyes, mouth, and, to a lesser extent, to the subtle intonations of their voice. Indeed the language of the eyes and the language of the lips are important means of stimulating the brain's trust circuits. And because it's easier to fake a trustworthy smile, the brain pays more attention to the involuntary movements of the muscles surrounding the eyes.

One type of gaze will attract us, while another will turn us away, and it only takes a split second for an astute observer to notice this physiological change. Thus from the perspective of neuroscience, the old axiom turns out to be true: when it comes to building trust, first impressions really count. If we see signs of happiness, our trust increases, but if we see the slightest bit of anger or fear, our trust will rapidly decrease.[1]

Honesty and Deceit

We need to look at a person's face to evaluate their trustworthiness, but the moment we realize that someone is looking at us, the brain shifts into a brief state of anxiety and alertness as it decides if the person is a friend or foe. Obviously, this presents a dilemma when it comes to first impressions, because it means that we are likely to see a face that appears anxious and thus untrustworthy—at least if the person is aware that we are looking at them.

This neurological problem is a reminder that first impressions really only provide a hint about a person's character and integrity. The same holds true when it comes to the notion of love at first sight. For example, the person you see sending you that special look of desire may, in fact, be thinking about someone or something else. You may think they're interested in you, but they're really enamored of the pastry displayed in the baker's window behind you. We can use first impressions as a clue, but we need to gather far more information as we engage the other person in conversation.

What causes us so much concern when we realize that someone is gazing at us? The philosopher Jean-Paul Sartre called it "the look," and he believed that the moment we realize someone is watching us, we become uncomfortably self-conscious. Neuroscience validates this premise to some extent but mostly for people who feel anxious or are behaving deceitfully. Eye contact tends to increase trustworthiness and encourage future cooperation among people engaged in positive social behavior.[2] The same effect is generated when we see someone with a happy gaze.[3]

Sartre argued that being gazed at can cause us to feel shame. In fact, he assumed that when we are alone our social morality fades away. Here the research supports Sartre's view. In a unique experiment designed by the Evolution and Behaviour Research Group at the University of Newcastle, researchers set up a coffee, tea, and milk station in an office. The price of each item was posted and an "honor" box was placed on the table for people to put their payments in. The researchers added one other element: a picture hung next to the price sign. For five of the experiment's ten weeks, different pictures of flowers were posted, but on alternating weeks, photographs of different pairs of eyes were posted so that they stared directly at the person who was standing in front of the beverage station. During those weeks, three times the amount of money was collected.

Clearly there was less cheating and more generosity when the office workers subconsciously perceived that they were being monitored, not by a person but by a photograph! As the researchers explained, "The human perceptual system contains neurons that respond selectively to stimuli involving faces and eyes, and it is therefore possible that the images exerted an automatic and unconscious effect on the participants' perception that they were being watched."[4]

In response to this research, The Newcastle police department launched an anticrime initiative in which it hung posters of glaring eyes, with the byline "We've got our eyes on criminals."[5] The result: crime dropped 17 percent during the first year after the posters were displayed around the town. A similar experiment has been running for several years in Derbyshire, England, in which cardboard cutouts of police officers were placed around town.[6] This did deter shoplifters and gas thieves but didn't seem to deter one type of thief: many of the police cutouts were stolen!

As other laboratory experiments have shown, people increase their levels of honesty and cooperation when they think they are being observed. But when anonymity is assured, we tend to act more selfishly, with greater dishonesty and deceit.[7]

The Language of the Eyes

Eye contact is a central element in social cognition, and everyone—from birth to death—depends on it to help them read other people's emotional states.[8] For infants, gazing into others' eyes is crucial for the neural development of the brain. It enhances cognition, attention, and memory, and it helps infants to regulate their emotions.[9]

Sustained eye contact initiates an "approach" reaction in the brain and signals that the parties are interested in having a social engagement.[10] But if one person averts their eyes, it signals an "avoidance" response to the viewer.[11] An averted gaze also sends a neurological clue to the observer that the person may be hiding something or lying.[12] But what that something is, we cannot discern unless we engage the person in a dialogue. For example, the person may feel a romantic attraction, but the fact that they are married can make the eye contact uncomfortable. Or perhaps the person is really busy and doesn't have time to initiate a social exchange. People with social anxiety will also avoid eye contact with others.[13] Eye contact is essential for the communication process, but the degree of contact can be influenced by the culture in which we're raised.[14] Thus we have to take many factors into consideration if we want to use our eyes to build conversational trust.

It's also not actually the eyes that communicate but the muscles surrounding them. If you pay specific attention to the movements of a person's eyelids and eyebrows, you'll receive vital information about their emotional state, especially about feelings of anger, sadness, fear, or contempt. Happiness and contentment are more difficult to discern, and a fully relaxed face can give the viewer the impression that you're not very interested in them.

Let's try a little experiment. Go to a mirror and take a few minutes to relax and breathe deeply. Scrunch up all the muscles in your face and then relax them. Do this several times and pay attention to the emotional messages that seem to be imparted. A tense face can convey anger, disgust, or disdain, but if you raise your eyebrows and open your mouth as wide as you can, depending on the muscles you use and how tight or relaxed they are, you can generate a variety of emotional messages, ranging from fear to surprise to terror.

Again relax your face muscles and just gaze at yourself for three or four minutes, paying attention to the thoughts and feelings that arise. If you begin to feel uncomfortable, stay with the exercise as you observe the feelings that come up. It should only take a few moments before the discomfort fades away.

Next see if you can consciously make faces that express the anger, sadness, and fear. If you use your memory of past events, you may discover that your face will reflect a deeper and more authentic emotional state. In fact, our emotional memories can stimulate the same muscular contractions that occurred when we experienced the real event.

Now experiment with different positive emotions: happiness, pleasure, contentment, peacefulness. Are they easier or harder to express? Again pay close attention to the inner speech that occurs with each expression you make. Finally, try emulating the expressions of shame, guilt, curiosity, boredom, and surprise. According to facial expression expert Paul Ekman, the more you feel the underlying emotions, the more you train your brain to both recognize and express them when you engage in a dialogue with others.[15]

Most of us are not aware of the expressions we convey to others, nor do we give must attention to the expressions on other peoples' faces. Thus we often mistake one emotion for another. But even an excellent reader of micro-expressions—emotional cues nonverbally communicated in less than a second—knows that these are just clues and that they have to be validated through deeper conversation. It's also important to remember that when a conversation becomes intense, there can be so many internal experiences going on that the messages our face imparts become blurred.[16]

We recommend that you do a similar experiment with your family members and friends. As in a game of charades, see if you can guess what emotional expression the other person is displaying. These exercises will help you to become more aware of the nonverbal messages we constantly send to one another. These exercises will also make you feel more comfortable with gazing intently at others while they are talking to you.

The Power of Gazing

There's one more experiment we'd like you to try with a partner, colleague, or friend. All you need to do is to gaze into each other's eyes for about five minutes. Most people will begin to feel uncomfortable after thirty seconds, but we want you to ignore the impulse to turn your eyes away. Instead sit with the discomfort and observe it, watching all the thoughts and feelings that are stirred up. Then take a few deep breaths and consciously relax your face, shoulders, and neck as you continue to gaze at your partner. When you've completed this exercise, talk about your experiences with each other.

This is a core exercise in our Compassionate Communication training program because it's essential to learn how to pay close attention to the other person's facial expressions throughout a conversation. We normally have people pair up with someone they don't know well and gaze at each other for a minute. Then we ask them to pair up with a different partner and try it again. Each time, it becomes easier, but it usually takes three or four rounds of gazing into different people's eyes before everyone in the group feels comfortable.

In order to succeed at this, you have to soften the muscles around your eyes. Otherwise your gaze will look more like a hardened stare. That type of gazing actually causes stress on the heart and will be perceived by others as a threat.[17] The result: the other person will avert their eyes, a signal that they are feeling discomfort. It's also a sign that trust is fading away.

There's another type of gaze, one that immediately stimulates a deep sense of trust and intimacy in the other person's brain. This gaze can't be faked because it involves involuntary muscles. The eyes are soft, and they reflect a sense of inner contentment and peace, but the expression also involves a specific type of smile, so let's take a few minutes to explore the language of the mouth. Then, toward the end of this chapter, we'll teach you how to generate a facial expression that reflects your inner feelings of cooperation and communication, an expression that will stimulate a deep level of trust in nearly anyone who observes your face.

The Language of the Lips

When it comes to developing empathetic trust, the eyes only tell part of the story. The other key facial expression concerns the mouth, for no matter how soft your eye gaze may be, even the slightest frown will convey a message of sadness or contempt, as the left-hand photo below illustrates. Fear is primarily communicated by the eye muscles,[18] but the slightest smile can neurologically communicate a feeling of peacefulness, contentment, and satisfaction, as the photo on the right demonstrates.

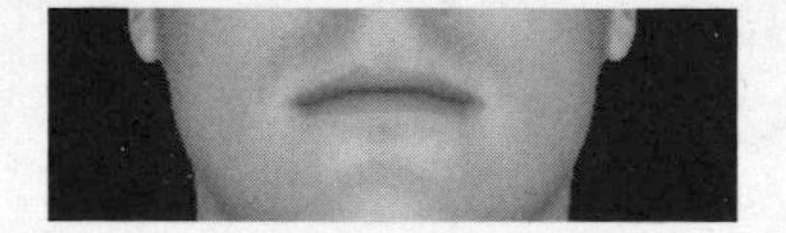

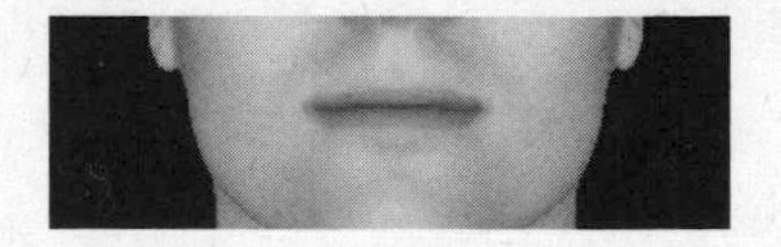

When we gaze at another person's face, the brain identifies a range of possible emotions in the eyes, and it does the same for the mouth. Indeed, as Ekman has documented, a single person can generate over ten thousand facial expressions, and many of them will trigger a specific neurological response in the observer's brain. With so many possibilities (and so little room in that window of consciousness known as working memory), the brain makes an educated guess as to what the other person is actually feeling.

The brain also looks for inconsistencies. If a person is lying or experiencing confusion, the eyes and the mouth can impart emotions that seem to conflict with each other. For example, in the picture above, on the left, the mouth could be expressing anger, sadness, or disgust. But if you combine that mouth with the eyes in the left-hand photo below, the emotion becomes clearer, conveying a sense of sternness.

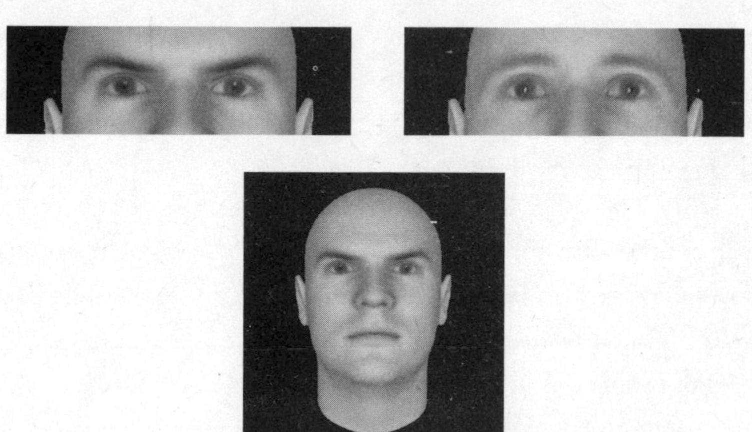

You'll still need other clues, like tone of voice, to tell if the person is irritated, contemptuous, or just concentrating very hard, but you can sense how the brain works to rapidly build an opinion—right or wrong—about what that individual is feeling or thinking.

The Language of Sadness

When it comes to generating neural resonance between two people, Ekman found that the most powerful facial expression is sadness. In fact, the more a person's face expresses suffering or pain, the more the compassion circuits are stimulated in the brain of someone looking at them. However, expressing sadness in front of another person often makes one feel vulnerable, so most of us cover up our hurt by putting on an expression of anger. It's a bad strategy, for as we've been documenting, anger tends to generate greater irritability, which leads to greater conflict. Thus it is always in our best interest to communicate our feelings of sadness and hurt and to suppress our defensive urge to get mad.

Just notice the feelings that are evoked when you look at the picture below. It should stir up many deep feelings, but anger won't be one of them. Interestingly, the brain tends to respond more compassionately when we see a child suffering, perhaps because we recognize the helplessness of the victim.

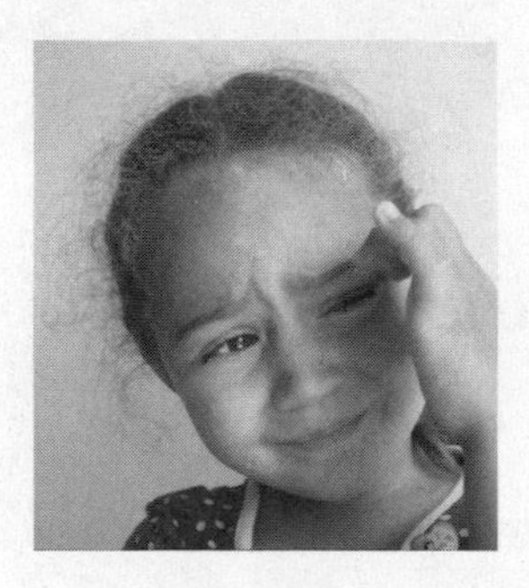

Ekman recommends that we train ourselves to express sadness, and it's rather easy to do. Just take a moment to remember a time when you felt particularly sad, and notice how the feeling affects the muscles around your eyes, mouth, and cheeks. Deliberately increase the feeling, letting it grow as strong as you can, and notice how it affects your thoughts.

Next stand in front of a mirror and see if you can duplicate the expression of the little girl above. Ekman suggests that you pull down the corners of your mouth and raise your cheeks as if you're squinting. Then look down, pull your eyebrows together, and let your eyelids droop.

Then, when you engage in conversations with other people, consciously try to mirror their facial expressions. As the research in these chapters has documented, the more you emulate the body movements and facial expressions of the person you're conversing with, the more your brain will resonate with theirs. You'll both feel more connected and empathetic, and this will generate deep trust.

The only emotion you don't want to mirror is the other person's anger. In this case we recommend that you turn your attention inward and focus on staying as relaxed and calm as possible. Use your imagination and immerse yourself in pleasant feelings or memories and try to generate as much kindness and compassion as you can for the person who is frustrated and mad. If you can't do this—if you feel your temper rising—call for a time-out and take a break, even if the other person objects. When they calm down, you can resume the conversation on a more positive note.

A Face Is a Face Is a Face

Can a robot use facial expressions and body language to win your trust? Yes, as researchers at MIT have proven with the creation of Nexi, the world's first social robot. When "she" moves her mechanical eyebrows, eyelids, and lower jaw, your brain will emotionally respond in much the same way as to a person.[19]

If you watch the video of Nexi at http://www.youtube.com/watch?v=aQS2zxmrrrA, you'll see how emotionally effective she can be.

The Million-Dollar Smile

A smile has enormous power: it can even change the electromagnetic activity of your brain.[20] But the ideal smile, as Leonardo da Vinci discovered, is really a half smile, because it enhances the quality of gently gazing eyes.[21]

A broad toothy smile will have the opposite effect. It often implies that the person is covering up anger or fear. Anxiety and irritability can make your jaw tense up, and if you try to smile, it appears forced.

It takes a special inner feeling, a feeling of genuine enjoyment, to generate a *Mona Lisa* smile. This "felt" smile, as researchers call it, can be elicited by a pleasurable experience, image, feeling, or thought, and when a person experiences this type of smile, their empathy toward others increases.[22] When you learn how to consciously generate and maintain this smile throughout the day, you'll feel more positive, your work will feel more pleasant,[23] and it will improve the demeanor of anyone you talk with because smiling has a contagious affect. [24] It will also strengthen the brain's ability to maintain a positive outlook on life.[25]

From the moment of birth, smiling, trust, and social empathy are neurologically entwined. When a mother sees a happy infant, dopamine is released in her brain's reward centers, and she smiles too.[26] Infants will also initiate a smile in order to communicate with a parent,[27] but if the mother is being inattentive, the smile will quickly fade away.[28] This is just another reminder that when we engage in conversations with others, we need to give them our fullest attention or their happiness will likewise fade away.

Gazing into the Eyes of the Beloved

Like many revelations in science, we accidentally discovered a way that will generate a Mona Lisa smile and gaze. In one of our Compassionate Communication seminars, Mark was preparing to guide the participants through the eye-gazing exercise described above. Normally, about 70 percent of the participants would begin to feel uncomfortable within the first minute.

Because the group was small, Mark decided to try something different. After everyone paired up, he had them close their eyes, and he guided them through a relaxation exercise. Then he asked them to think about someone they deeply loved, or a memory that brought them a deep sense of pleasure and satisfaction, visualizing it with as much detail as possible. Within a matter of seconds, everyone in the room seemed to radiate a blissful expression, the smile that Leonardo captured so beautifully in his painting.

When the participants opened their eyes to gaze at the person they were facing, the smiles seemed to grow more intense, and when Mark asked them to talk about their experiences, everyone spoke slowly and softly. They all agreed that they felt genuinely cared for by the other person, even though they had never met before.

The following month, Mark guided a group of 110 people through a similar exercise. First he just asked them to gaze at each other for thirty seconds, without any preparation. When he asked how many people felt discomfort, about three-quarters of the participants raised their hands. He asked them to pair up with someone different, and then guided them through the visualization exercise. The same soft smile emerged on nearly everyone's face.

Mark asked the group to open up their eyes and to gaze at the other person for two minutes. How many people felt uncomfortable with the gazing? Only four hands were raised.

A 2010 brain-scan study validated this anecdotal evidence. At the Institute of Neuroscience in Taiwan, researchers discovered that imagining a loved one promotes greater empathy and compassion for others by stimulating activity in the anterior cingulate and the insula.[29] The same areas are stimulated when mothers view their smiling infants and when people engage in loving-kindness meditations.[30]

Give it a try, right now. Take a few deep breaths and bring yourself into the present moment. Relax all the muscles in your face, your jaw, your neck, your shoulders, and your arms. Take a few more deep breaths and think about someone you deeply care for or recall an event in your life that brought you deep satisfaction and joy. Imagine that you're right there, with that special person, or in that special place, and feel how that *Mona Lisa* smile begins to light up your face.

Now take that smile into the world and share it with as many people as you can.

PART 2

The Strategies

Developing New Communication Skills

CHAPTER 7

Inner Values

The Foundation of Conscious Living

"No" may be one of the most powerful words in the world, but it's not necessarily the most powerful word in *your* life. You'll have to discover for yourself what the most powerful word in your life is, but it will be a word that encapsulates the most important principle in your life. And when it informs the other words you say, it will both protect you from being knocked off balance in verbal conflicts and help you to stay focused on achieving your personal and professional goals.

The question that will help you identify this word is an essential one in every person's life, and yet one we rarely ask ourselves. In fact, it's so rare that if you enter any variation of it as a Google search, it brings up fewer than fifty results. Yet a question like "What makes me happy?" can bring up as many as twenty-eight million hits.

For this exercise, we'd like you to have a pen and a piece of paper handy, and as we've done in most of the previous exercises, to start by taking a few minutes to ground yourself. When you feel fully relaxed, ask yourself: what is my deepest, innermost value?

Close your eyes for at least sixty seconds, listening to your inner voices and paying attention to whatever thoughts and feelings float through your mind. Then open your eyes and write down a single word or brief phrase that captures your deepest value.

If nothing occurs to you, close your eyes again and stay focused on the question for another couple of minutes until a word comes to mind. Write it down, and repeat the question: what is my deepest, innermost value? If a different word comes to mind—and it often does—write that one down as well. Repeat this step several more times, to see if other essential values rise into consciousness.

Now look at your list of words, and circle the one that feels the truest for you at this moment. Close your eyes once more and repeat the word or phrase to yourself, silently and then aloud. Notice how it feels to say it, and then compare it to the other words you wrote down.

What is the point of doing such an exercise? According to researchers at the University of California, Los Angeles, "Reflecting on personal values can keep neuroendocrine and psychological responses to stress at low levels."[1] This is truly amazing: by simply pondering and affirming your deepest values you'll improve the health of your brain, you'll protect yourself from burnout at work, you'll reduce your propensity to ruminate about failure, and you'll be less reactive and defensive when someone confronts you with uncomfortable information.[2]

The Ten-Day Experiment

Try doing this "inner values" exercise for the next ten days. It's the first assignment Mark gives to his students on the first day of class in the Executive MBA Program at Loyola Marymount University, Los Angeles, one of the top-ranked business schools in the world. This specialized program is designed for full-time managers, executives, and business leaders who need to learn advanced skills for maintaining a growing and successful company.

Here's what we'd like you to do. Each morning, shortly after you wake up, take a few moments to stretch, breathe deeply, and relax. Then ask yourself, what is my deepest, innermost value? Create a log and record your words, along with any feelings or reactions you have relating to doing the exercise. Do this for ten days, and on day eleven briefly answer the following seven questions, using only a single sheet of paper. Be spontaneous in your responses, and remember that there are no right or wrong answers to these questions. They're only designed to deepen the self reflective process.

1. What was your initial reaction to this exercise?
2. Was the exercise enjoyable, boring, interesting, annoying, etc.?
3. How long did you spend, each day, contemplating your inner values?
4. Did the exercise have any effect on other aspects of your day, work, or life?
5. How do you define the word "value"?
6. Did you discover anything about yourself?
7. Did the exercise influence the way you think about your work and business values?

In Mark's class this homework assignment was optional, and the students who completed it were asked to submit their daily logs, along with the answers to the above questions, anonymously. He didn't ask them for their names because he really wanted to know if the exercise had any immediate or lasting value for the busy executives enrolled in the program.

In the end nearly everyone found the exercise useful, enlightening, and enjoyable, but it didn't start out that way. Some were intrigued, others were bored, and a few actually became irritated with the assignment. One student—a chief operating officer at a midsize corporation—put it bluntly: "What the *#!* does this have to do with financial planning?" But by the end of the ten days, he wrote the following comment: "I think that this exercise should be taught to every MBA student in America." He was not alone, as the following excerpts illustrate, taken from the students' written responses to the seven questions listed above:

> At first I thought, "Who has time for this?" I barely have enough minutes in the day to run my company, and the workload for the MBA class is overwhelming. But those couple of minutes each morning helped me stay calm and focused for the rest of the day. I plan to do this exercise for the rest of the school year.

> The moment I awake, my mind rushes to plan the day. This exercise made me realize that I'm undermining my health. I get the most from it when I practice five to ten minutes a day, and I've noticed that the quality and quantity of my sleep has improved. I know I have strong values, but I've never taken the time to acknowledge it.

> I really became more conscious about my emotions, and how they could sabotage my evenings with my wife. Once, after having a fight with her, I spent thirty minutes sitting alone, thinking about the value of my marriage. I went back and apologized, and we worked our problem out.

> I used my positive word all day long. I felt calmer, less stressed, and it seemed to help when it came to solving difficult problems at work. I loved the self-awareness it brought, and the way it made me feel throughout the day.

> The core values that kept coming up for me were honesty, integrity, and family. It made me think about my business ethics and values, and what was really essential for work. I realized that I'd rather climb the ladder of success more slowly so I can support the people I meet along the way, and give more time to my family.
>
> This exercise grounded me in the principles of goodness and the desire to live by my deepest principles. For me, work can drown out the self-talk of my core values. When that happens, I can't truly express who I am or realize my greatest potential.
>
> At first I hated this exercise, but it forced me to reexamine my priorities. I realized that business is not just about numbers and money. I think everyone needs to find at least two minutes a day to think about their values and principles and how to use them to build a life-sustaining career and personality.

More than a third of the students said that the exercise inspired them to become more involved in spiritual pursuits like meditation, even though no mention had been made of them. But even more surprising, several people wrote that they were going to restructure their companies to be more values oriented. One CEO asked every member of his company to write up a personal "mission and values" statement, which he collated and distributed to the class.

Now, it may be just a coincidence, but the couple of students who ignored the exercise had greater difficulties with their schoolwork. And when they were later involved in teamwork activities with their classmates, they tended to be less cooperative and more stubborn.

Inner Values on the Internet

Over the past two years, we've been able—using Facebook and other social media forums—to get feedback about this exercise from people all over the world: college students, therapists, religious practitioners, divorce attorneys, mediators, teachers, corporate executives. And the feedback is overwhelmingly positive—perhaps because it takes so little time to do. For example, here's what happened when John, a construction worker from New Zealand, did the "inner values" exercise for ten days:

> My initial reaction, was, "Not another thing to do!" But then I realized that I hadn't put much focus on values in the past, even though I'd read about it. Love, service, and family were my three top values, and I started to realize where love was really missing: at work. Normally, I feel a lot of animosity toward my boss, but by the third day of my values experiment, I started feeling kindness toward him. I began to let go of my anger because I saw that he was only doing his job. Then I started feeling gratitude, because he was the one who gave me my job.

Cheri Frootko, a South African film director and script supervisor saw the inner values exercise on YouTube, in a clip from a TEDx talk in Thousand Oaks, California, that Mark gave in 2010 (you can see the TEDx talk at http://www.youtube.com/watch?v=yvhCLXEeSDQ). Cheri had just assembled a team to shoot a project in France and decided to show them the video. Each morning, before they began work, they practiced the inner values exercise:

> We created a fun routine. We imitated Mark on the video: yawning, breathing deeply, stretching, rolling the shoulders, and shaking

our hands. We closed our eyes and asked ourselves what our greatest value was, and then, in a spirit of lightheartedness, we shared our words with each other. The result? Ten people, who a week earlier were total strangers, created a bond of insight and intimacy. And it wouldn't have happened without this three-minute catalyst. We would have worked well without the exercise—say, on a level of 6—but with our sharing of values, the energy and harmony of the group reached a level of 9. PS: Forgot to mention that when the pressure got intense, we used a specific buzz word on the set—"yawn"! It made everybody relax and lighten up.

Exercises like this are slowly working their way into business and medical communities. At Missouri State University, psychologists found that when a personal values exercise was included in a treatment plan designed to help patients cope with chronic pain, their tolerance toward pain improved.[3] When we get in touch with what is most meaningful in our lives, we are less distracted by the problems that occur throughout the day.

Are We Moving Toward a Values-Based Society?

Inner values used to be a popular topic in the 1950s and 1960s, when books by Viktor Frankl (*Man's Search for Meaning*) and Abraham Maslow (*Religion, Values, and Peak-Experiences*) were best sellers. But during the past twenty years, values-based research mostly disappeared.

Recently the picture has changed. With the meltdown of the financial institutions that occurred several years ago, magazines like *Bloomberg Businessweek* have been regularly calling for the implementation of corporate and leadership values. And the business world is responding.

Harvard business professor Rosabeth Moss Kanter—considered by many to be one of the most powerful women in the world—recently commented on the importance of directly addressing values in the boardroom: "In organizations that I call 'supercorps'—companies that are innovative, profitable, and responsible—widespread dialogue about the interpretation and application of values enhances accountability, collaboration, and initiative."[4]

Our own research supports this. Even though everyone has a unique set of values—running the spectrum from highly idealistic principles like truth, integrity, and growth to highly interpersonal values like love, family, and friendship—when people openly share their values with each other, they come together and express mutual support.

We once had a church auditorium filled with religious believers and disbelievers, liberals and conservatives, millionaires and welfare recipients, and when we guided them through the inner values exercise, and then asked them to share their values aloud, nearly everyone (as far as we could tell) ended up feeling a deep sense of mutual respect for each other. And when the president of the largest atheist association in America told the group that his inner value was to help people find deep peace in their personal and professional lives, everyone in the room applauded.

Kanter finds that the same thing happens in the business world. When people share and discuss their deepest values, it strengthens the motivation of the entire group. Employees' personal values become integrated with the company's policy, and this helps guide the ethical choices of the corporation. By discussing business values openly, Kanter argues, it eliminates the need to impose impersonal and coercive rules.

In Kanter's experience, discussions about values also help to decrease interpersonal conflict. Cooperation grows, everyone feels like they are part of the team, and profitability increases for everyone:

> The organization becomes a community united by shared purpose, which reinforces teamwork and collaboration. People can be more readily relied on to do the right thing, and to guide their colleagues to do the same, once they buy into and internalize core principles . . . And, as I have seen in leading companies, active consideration of core values and purpose can unlock creative potential.[5]

That is the power of a single question.

What, Exactly, Is a Value?

We are often asked to define what we mean by "value." But the beauty of this exercise is that we *don't* define it. Nor do we give examples. When someone else suggests what values we should consider, the exercise becomes outer directed, not inner directed. If people to ponder the question in their own ways, remarkable self-discoveries can be made.

Values are difficult to define or categorize because they can touch on so many dimensions of life. There are moral values, political values, religious values, marital values, organizational values, and aesthetic values. There are practical values and theoretical values, scientific values and philosophical values. There are personal values and interpersonal values, health values and money values. Values can even govern the types of food we eat and the products we buy.[6] But if they become too rigid—or turned into "shoulds"—they can generate a myriad of conflicts with others.[7]

Inner values are shaped by both genetic and environmental influences,[8] and they are essential for providing meaning and purpose to life. Without them, we're more inclined to exhibit antisocial behavior.[9]

Interestingly, different values activate different structures within the brain,[10] and it's even been shown that people with different cultural values activate different areas in the visual cortex.[11] They may actually see the world in a fundamentally different way.

Asking the Right Questions

When we were first gathering data to measure the effects of Compassionate Communication, we asked workshop participants the following question: what is your secret desire? The question was inspired by the phenomenal success of the movie and book *The Secret*, and we were curious to see how people would respond. It gave us a goldmine of valuable information.

Before participants were guided through the full Compassionate Communication script, most people responded to the "secret desire" question with materialistic goals: more money, a better job, a nicer house, etc. After practicing the dialogue exercise for forty minutes, people responded very differently. Happiness and contentment were often cited. Financial desires dropped from 34 percent to 14 percent, whereas a yearning for peace increased by 60 percent. Desires for self-love and interpersonal love nearly tripled.

Intrinsic values like these are far more likely to be associated with satisfaction in life and emotional well-being than with wealth.[12] That's why it's important to ask the right question, in the right way. If you ask people what they want, their answers will often focus on material prosperity. But if you ask them what makes them happy, money is rarely mentioned. Happiness, it turns out, is a universal value that is far more important to people than material wealth.[13] Money may be desirable, but it cannot buy you trust or help you develop positive emotions—elements that are essential for achieving satisfaction. In his book *The Social Animal*, David Brooks argues that a person who is happy in their job but has a bad family and social life is much worse off than someone who struggles at work but has a great family life.

In reality, the stress created by focusing too much on money can literally threaten our lives. To quote a 2010 study conducted at the University of Liège, "Money impairs people's ability to savor everyday positive emotions and experiences."[14] The study found that wealthier individuals had far more difficulty enjoying their lives than people who earned moderate amounts of income.

Situational Values

As you experiment with the inner values exercise, you'll find that the results will change and evolve over time. Specific events—such as marriage, divorce, or becoming a parent—can dramatically alter our values, for better or for worse. For example, a nasty divorce may cause a child to view marriage suspiciously, but it can liberate a spouse to find a partner who shares similar values and beliefs.

Surprisingly, a life-threatening event will make most people revise their values in ways that bring them greater satisfaction.[15] Even reflecting on death tends to shift values away from greed and toward unselfish and caring behavior.[16]

In our research there are two variations of the inner values exercise that people have found to be useful. If you take a moment to reflect on the following questions, you'll see that the answers are different from the ones relating to innermost values: what is my deepest relationship value, and what is my deepest communication value?

Most people have similar responses to these questions. For the relationship value, the most common words chosen are "kindness" and "trust." For the communication value, it's the desire to be listened to with respect and to be spoken to with honesty and warmth. If we consciously exercise these values whenever we engage in dialogue, the odds of conflict are strongly reduced, even when we are interacting with people we dislike or distrust.

I'll give you an example of what occurred when Mark was called into an executive board meeting to negotiate a heated dispute. The organization was a psychology training center, and the issue concerned the expression of anger. One group of therapists believed that the honest expression of anger was essential for the healing process. The other group, comprised mostly of the corporate leadership staff believed that tact and diplomacy were paramount.

The leaders of the two factions could not compromise, and a stalemate had been reached. Mark asked each leader what their innermost personal value was, what their deepest relationship value was, and what their deepest communication value was. Sam, the proponent of tactfulness, went first, and his three words were "love," "compassion," and "gentleness." Jill, who strongly believed in the need to "get the garbage out," as she referred to the psychological theory of emotional release, had a somewhat different list: "kindness," "integrity," and "honesty."

"Perfect!" Mark said. "Do you both respect the other person's values?"

They nodded in approval.

"Then, Jill, I want you to continue your argument with Sam, but you have to honor both Sam's and your own sets of values. I want you to express your anger honestly but with love, compassion, kindness, and gentleness."

She couldn't do it, because it's impossible to express anger, resentment, or for that matter any negative emotion, in a kind and productive way. Two months later, Jill resigned, and the company has continued to flourish.

Inner Values in Personal Relationships

When people share their personal, relationship, and communication values with each other *before* discussing a difficult issue, they are more likely to remain emotionally calm and centered. Such discussions have been shown to be particularly useful for improving communication in couples counseling, because it derails feelings of anger, distrust, and contempt before they creep into the dialogue. Here's an example of how James Walton, Ph.D., a licensed marriage and family therapist in Los Angeles, uses the inner values exercise in his practice:

> With my patients, I have them contemplate their deepest value for two minutes every day between sessions. Those that have done so have experienced some amazing transformations. When working with couples, I have them explore the role that personal values play in the relationship and how, if we violate those values, it creates problems. With my clients, love and compassion are the most common values reported.
>
> Let me give you an example. Clara and Bart were having multiple problems communicating with each other. He was passive-aggressive in his behavior toward her and she was openly critical and hostile toward him. They would fight over small things, which would lead to hostilities that were out of proportion to the infraction. They had stopped being friends in the relationship, so I decided to try the inner values exercise with them.
>
> I had them do a brief relaxation exercise; then I asked them to visualize in their minds someone they loved dearly and to feel those feelings of love. I then asked them to focus on their greatest core value and to allow it to come forward in their mind. For Bart, it was his need to feel supported. For Clara, it was her need to feel accepted.

When we discussed their experience, they realized that they were not being sensitive to the other person's core value. I explained that when we violate our own core value, we feel disempowered. I helped Clara become aware of how important support was to Bart, and I asked Bart to practice meditating on the essence of support for a few minutes each day, using the inner values technique. He was to focus on what it felt like receiving support, giving support, and filling his heart with that experience. I also asked Bart to think about how he could show Clara more acceptance.

Clara was asked to meditate a few minutes each day on the concept of acceptance—to feel acceptance in her heart for herself and for others, and to feel what it was like to give and receive it.

One week of practice made a dramatic improvement in their relationship. They each said that they felt much closer to the other and more understood. They were becoming friends again, and this one exercise did more for their relationship than all the other work we had done before.

I have employed this technique with other couples, and in each case it has helped them. If they practice daily, contemplating their greatest value, the results are dramatically better, because it helps both individuals feel more empathy for each other.

Is there a general rule that incorporates the most basic values of communication? We think so. It is a paraphrasing of the Golden Rule: speak unto others as you would like them to speak unto you, and listen to others as you would like them to listen to you.

Establishing Lasting Business and Professional Values

We all seem to share similar communication values, but research is beginning to show that personal and professional values frequently differ.[17] This can present a problem, because when there is incongruence between inner values and work-related values, emotional burnout is likely to take place.[18]

In the health-care and medical community, this happens frequently. For example, physician burnout has been estimated to be close to 50 percent in some parts of the country, and a study of thirty-two hundred Canadian doctors could actually predict who would experience exhaustion and poor work performance by identifying the people whose personal values conflicted with the values promoted in the work environment.[19]

This has strong implications for the business world. As researchers at the University of California, Los Angeles, advise, when your management strategies match the value of your workers, greater job satisfaction is reported, and less people are likely to quit.[20]

Here's an exercise created by Dr. Roger P. Levin, DDS, which anyone can use to integrate their personal and professional values:

> To identify business values, make a list of approximately 15 words that you feel are the core of your practice [i.e., business] values or beliefs. This list might include such terms as integrity, balance, profit, growth, challenge, caring, excellence, quality, trust, appreciation and enthusiasm. After you have created a list, the key is to spend the next 10 days paring it down to no more than six words. The rule is that you can add a word to the list, but only if you take one off. You can combine words that have similar meaning, such as integrity and honesty. You ultimately will have to eliminate less important words . . .

> Once you know your four to six business values, you can strengthen your practice and build a high-powered team . . . [then] repeat the process for your personal values. It can be insightful and fun.[21]

Spiritual Values

Throughout most of history, the question of values has been a spiritual one, and sacred texts have attempted to identify which values will lead to the greatest satisfaction, in this life and whatever may lie beyond. But if we were to make a list of all the spiritual values that have been proposed, we could probably fill this book from cover to cover.

Despite centuries of theological debate, our species has yet to come up with a mutually agreed-upon list of which values are most essential for our happiness or our survival. Yet everyone has an opinion. Perhaps the easiest way to explain this lack of resolution is to compare it to the nature of the human brain. Unlike other animals, each person has a unique pattern of brain activity, and as we've explained previously, no two people—and no two brains—give the same meaning or value to the same word. We're unique, and so the values we choose to live by, to speak through, are as unique as the ever-changing neurons that shape the decisions we make.

Even the word "spirituality" has defied definition within the religious, philosophical, and psychological communities. But to our way of thinking, spirituality and values are often the same. We choose to ask about your "innermost" value rather than what your "highest" value because this phrasing sidesteps theology and speaks equally to believers and disbelievers alike. An ongoing research project at the University of California, Los Angeles, takes a similar approach:

> Spirituality points to our interiors, our subjective life, as contrasted to the objective domain of material events and objects. Our spirituality is reflected in the values and ideals that we hold most dear, our sense of who we are and where we come from, our beliefs about why we are here—the meaning and purpose we see in our lives—and our connectedness to each other and to the world around us. Spirituality also captures those aspects of our experience that are not easy to define or talk about, such as inspiration, creativity, the mysterious, the sacred, and the mystical. Within this very broad perspective, we believe spirituality is a universal impulse and reality.[22]

In this world of competing beliefs, we feel it is essential to promote a values-driven dialogue that while related to political and religious beliefs for many people, also transcends those beliefs. Thus the foundational element of Compassionate Communication is to honor the core values of both the listener and the speaker. All we have to do is to stop outside of the meeting room, or pause for a moment before we walk through the door of our home, and ask ourselves this question: what do I value most about the person I am about to meet?

If we did this more often, the risk of engaging in conflict would recede.

CHAPTER 8

Twelve Steps to Intimacy, Cooperation, and Trust

Speaking briefly. Speaking slowly. Listening deeply. Showing appreciation and remaining positive. Observing our inner speech and cultivating inner silence. Studying the other person's facial expressions, body gestures, and vocal inflections and mirroring them to build neural resonance. Focusing on your inner values and bringing them into every conversation as you remain as relaxed and as present as you can. These are the twelve strategies that are essential if you want to build meaningful, trustworthy, and long-term productive relationships with others. If you ignore any of them, the research suggests that you will compromise your ability to communicate and increase the risk of conflict.

Whether we are talking to a friend or a lover or a colleague at work, and whether we are talking to a child, a stranger, or a person suffering from an emotional or cognitive disease, these communication strategies will ensure the best dialogue possible. When we choose our words carefully, and orchestrate them with the strategies above, we enhance the comprehension of the listener in a way that fosters compassion and increases friendly cooperation. But the words we speak and listen to are only a small part of the communication process. It is the *way* we say them and the *way* we listen to them that makes all the difference in the world.

To improve our conversational skills we have to do several things. First we need to recognize that the way we normally speak is inadequate, filled with habituated patterns that were mostly set in place in adolescence and early adulthood. Then we have to consciously interrupt those speaking and listening habits, over and over again. And finally we need to replace those old communication styles with new and effective ones. This requires experiential training, and training takes time.

Fortunately, the twenty-minute exercise we explain in the next chapter will guide you through these twelve strategies and allow you to practice them with a partner. Even a few rounds of practice will be sufficient to give you enough experience to take these strategies and incorporate them into your conversations at home and at work. They will significantly improve your ability to empathize with others, and, according to our research studies of similar types of exercises, you should be able to alter the structure and function of key areas in your brain that relate to improved social awareness, enhanced cognition, and greater emotional control in eight weeks or less. You'll be actually rewiring your brain to communicate more effectively with others.

The Twelve Components of Compassionate Communication

In this chapter we'll review the evidence supporting each of the strategies that we want you practice when talking and listening to others. The first six steps are preparatory. They're what you do before you enter a room to engage another person in a conversation, and they are best carried out in the following order:

1. Relax
2. Stay present
3. Cultivate inner silence
4. Increase positivity
5. Reflect on your deepest values
6. Access a pleasant memory

These steps create an inner state of intense awareness and calm, which is essential for engaging in one of the most crucial aspects of communication:

7. Observe nonverbal cues

If you are not conscious of the subtle changes in the other person's tone of voice, facial expressions, and body gestures, you are likely to miss important clues that tell you what that person is really thinking and feeling. You won't know if the person understands you or if they're even paying attention to what you say. Then, when you engage in dialogue, the following five strategies should be consistently adhered to:

8. Express appreciation
9. Speak warmly
10. Speak slowly
11. Speak briefly
12. Listen deeply

How many people conscientiously apply these techniques on a daily basis? Far fewer than we would wish. It's like weight loss: we all know what's required, but we easily slip back into our old habits. It's human nature, and it takes a lot of neural energy to interrupt an old behavior. To build a new habit, we have to repeat a new behavior hundreds and hundreds of times. Eventually, it will become second nature. It begins by taking a few deep breaths and relaxing as you consciously bring your fullest attention and awareness into the present moment.

Step 1: Relax

Stress is now considered the number one killer in the world. Stress generates irritability, irritability generates anger, and anger shuts down the ability to communicate and cooperate with others.[1] So before you enter a conversation with anyone, spend sixty seconds doing any variation of the following relaxation exercises.

First notice which parts of your body are tense. Assign a number on a scale of one to ten (with ten being extremely tense) to signify your state of relaxation or stress. Write down the number on a sheet of paper.

For the next thirty seconds, breathe in slowly to the count of five, and then exhale slowly to the count of five. Repeat this three times. Now, if possible, yawn a few times and notice if your level of relaxation has increased. Assign it a number between one and ten and write it down.

Now slowly stretch your body in any way that feels comfortable and pleasurable, and see if you can immerse yourself completely in the sensation of each stretch. Begin with the muscles of your face, scrunching them up, then stretching them out. Then move down to your shoulders and neck, gently moving your head from side to side and from front to back. Scrunch your shoulders to your ears and let them drop, pushing them down toward the floor.

Next tighten up all of the muscles in your arms and legs. Hold them tightly as you count to ten; then relax them as you shake your hands and feet. Take a few more deep breaths and rest. Once more assign a number to your state of relaxation and write it down, noticing how much you've improved.

Can a brief exercise like this really change your brain in ways that will measurably improve your communication skills? Yes! Several fMRI studies have shown that a one-minute relaxation exercise will increase activity in different areas of the cortex that are essential for language, communication, social awareness, mood regulation, and decision making.[2] If you increase the length of this relaxation exercise, additional parts of the brain will be activated that help you become more focused and attentive at work.[3] Cortisol levels will drop, which means that your levels of biological stress will have decreased.

Research also shows that just watching the patterns of your natural breathing will change your brain in positive ways, and if you coordinate your breathing with another person, it will help the two of you to feel more calm and caring toward each other.[4] At the end of his book *Emotions Revealed*, Paul Ekman (the facial expression expert) writes, "I previously couldn't understand why focusing our awareness on breathing would benefit emotional life." But then, "like the proverbial bolt out of the blue," he explains, he had an insight:

> The very practice of learning to focus attention on an automatic process that required no conscious monitoring creates the capacity to be attentive to other automatic processes . . . We develop new neural pathways that allow us to do it. And here is the punch line: these skills transfer to other automatic processes—benefiting emotional behavior awareness and eventually, in some people, impulse awareness.

In conversations that get heated, the person who is capable of remaining calm will benefit the most. So by all means learn how to focus on your relaxation and breathing when difficult issues are discussed.

Step 2: Stay Present

When you focus intently on your breathing and relaxation, you pull your attention into the present moment. When we become completely absorbed in something as simple as breathing or relaxing a specific part of our body, the inner speech of everyday consciousness stops, at least momentarily, and this allows us to become aware of the subtle things that are immediately happening around us. We hear sounds we rarely notice, we feel more sensations in our body, and if we bring this "presentness" into a conversation, we hear more clearly the subtle tones of voice that give emotional meaning to the speaker's words.

Here's a little exercise created by the renowned author and spiritual teacher Eckhart Tolle that you can do right now to experience the power of the present moment.[5] Begin by concentrating on your right hand; then ask yourself this question: how can I really know, at this very moment, that my hand exists? The more you think about this question as you focus on your hand, the more sensations you'll begin to feel.

If you don't feel any difference after a minute or two, close your hand, very slowly, into a fist, and hold it tight for thirty seconds. Then spend the next thirty seconds slowly opening it back up. Pay attention to every sensation in each finger and your palm.

You'll notice that in this deep state of concentration, your mind has become silent, and although you may not realize it, your blood pressure will have dropped. Being relaxed and in the present moment is beneficial to your heart.

Keep focusing on your hand and compare its "aliveness," as Tolle calls it, to your other hand. Using this technique, you can bring that enhanced awareness to every part of your body, whenever you choose. You can also bring it into the conversations you have with others.

Using fMRI technology we can actually watch how this moment-to-moment awareness of the inner and outer world alters the functioning of the brain. Our everyday consciousness shifts into a meta-awareness that allows us to experience a larger and more unified perception of the world.[6]

If we bring this moment-to-moment awareness into our conversations with others, we will experience the interaction with greater clarity, and we'll be less likely to be knocked off balance by the other person's emotional state. We'll feel their pain and respond with compassion because we have been able to remain relaxed.

Being in the present moment has an interesting side effect: because you're less likely to control the direction of the conversation, it can lead to unexpected dialogues. If sadness comes up for you or the other person, and you remain in the present moment, the conversation will focus on those feelings and the previous topic will fade away. It's a very intimate experience, and thus very appropriate for conversations with family members and friends, but in business it's essential to stay focused on the specific topic of discussion. Being in the present moment, however, will allow you to quickly recognize when a conversation begins to go astray.

Step 3: Cultivate Inner Silence

Most of us are only able to stay relaxed and in the present moment for brief periods of time. Soon it gets interrupted by our inner speech. Research shows that you can suppress those distracting feelings and thoughts, but you have to practice doing it over and over until you gain control.

The more you consciously think about *not* thinking—as a formal training exercise—the more you gain voluntary control over the brain's spontaneous cascade of inner speech and cognition.[7] As researchers as Emory University found, thought suppression can even protect the brain "and reduce the cognitive decline associated with normal aging."[8]

We specially need to develop the skill to remain silent so that we can give our fullest attention to what other people say. Unconsciously they will know when we're distracted by our inner speech, and the lack of interest they perceive will make them distance themselves from you. Thus in active communication silence is not the enemy. It's your friend.

For many people, learning how to remain in a state of inner silence can be difficult because the temporal lobes of the brain are designed to constantly listen for something. And something is always making some degree of sound.

Here's a technique that we and other teachers use to show people how to cultivate a deeper state of silence. You'll need a bell that when rung will resonate for at least fifteen to thirty seconds. If you go to http://www.mindfulnessdc.org/bell/index.html, you can activate an online mindfulness bell that is perfect for this exercise. Push the button to "strike" the bell, then focus intensely on the sound. As the tone fades, you'll notice that you have to give more attention to your listening. Then, when the sound disappears, continue to listen deeply to the silence, which, as you will discover, is filled with a variety of subtle sounds. You might even become aware of the sound of your breathing, and this is an excellent sound to focus on (it provides substantial benefits to your brain).

Ring the bell again, and listen even more closely than you did before. Continue several more times as you train yourself to recognize the special state of awareness it puts you in. This is the state of attentiveness that we would like you to use when listening to another person speak. The online mindfulness bell will also aid you in the practice of the Compassionate Communication training exercise described in the following chapter.

Step 4: Increase Positivity

Before you begin any conversation, take a mental inventory of your mood. Are you feeling happy or depressed, tired or alert, anxious or calm? Any negative thought or feeling you have interferes with the parts of your brain that are involved with language processing, listening, and speech.

Research shows that the three previous steps are usually sufficient to eliminate negative feelings and thoughts. But if they still remain, consider the following choices: repeat the exercises above, or consider postponing the meeting, especially if it's related to work. When a colleague, employer, or employee senses your exhaustion or stress, they will know that your ability to have a meaningful, productive dialogue is compromised. So why take the risk?

Even if you feel calm and relaxed, ask yourself this question: do I feel optimistic about this meeting and the person I'm about to converse with? If the answer is no—if you harbor any significant degree of doubt, anxiety, frustration, or even an inkling of anger—then again, if possible, you should postpone your dialogue until a later date. If you can't postpone the dialogue, at least spend a few moments focusing on a more positive idea, because any negative state can generate mutual defensiveness and distrust.

Here's something to do when you have concerns about an upcoming meeting. Mentally rehearse what you think could happen. Have an imaginary conversation with the person you want to talk to, as if you were an actor reading from a script, and see where the dialogue goes. When you do this, it is easy to spot statements you might make that would undermine your intention and goal.

If you still feel upset or worried, then take the fantasy conversation to the next level and imagine how the other person might respond if you told them how you really felt at the moment. If it doesn't make them smile or bring a tear to their eye—if it doesn't make them feel like you respect them—then you'll know ahead of time that the conversation will likely fail.

To make any conversation truly satisfying and successful, you need to generate heartfelt positivity, for yourself and the other person. As Barbara Fredrickson, a distinguished professor of psychology at the University of North Carolina, says, positivity is our birthright,

> And it comes in many forms and flavors. Think of the times you feel connected to others and loved; when you feel playful, creative, or silly; when you feel blessed and at one with your surrounding; when your soul is stirred by the sheer beauty of existence; or when you feel energized and excited by a new idea or hobby. Positivity reigns whenever positive emotions—like love, joy, gratitude, serenity, interest, and inspiration—touch and open your heart.[9]

Fredrickson identified one of the most important factors for predicting success in both personal and business relationships. It's called the three-to-one ratio, and it's a comparison of the number of positive thoughts and negative thoughts you generate when you engage in a conversation with someone else. If you express fewer than three positive thoughts or behaviors for each negative one, the relationship or interaction is likely to fail. This finding correlates with Marcial Losada's research with corporate teams[10] and John Gottman's research with married couples.[11]

Fredrickson, Losada, and Gottman realized that if you want your business and your personal relationships to flourish, you'll need to increase your ratio by generating at least five positive messages for each negative utterance you make (for example, "I'm disappointed" or "That's not what I had hoped for" count as expressions of negativity, as does a frown or gesture of contempt). Someone with a positivity ratio that falls below three-to-one is likely be diagnosed with depression.[12]

We suggest that in preparation for a serious dialogue you use your imagination to visualize and rehearse a conversation that is filled with positivity, kindness, and optimism. As researchers at Purdue University found, when you enter a conversation with optimism both you and the listener will likely be more satisfied with the interaction.[13] And if you consciously visualize a future success, it will enhance your motivation to achieve it.[14]

The research is substantial: positive imagery can reduce a negative state of mind, whereas negative images will maintain or enhance a negative mood.[15] In fact, positive mental imagery, when compared to other forms of verbal processing, has a greater impact on reducing anxiety.[16] Negative imagery, however, will amplify it.[17]

This raises an interesting question: can you arbitrarily create an optimistic attitude by manipulating your own thoughts? Researchers at the University of Toledo say yes,[18] and you can even undo negative memories from childhood by "rescripting" the event and imagining a different outcome or solution.[19] So by all means, prime yourself with positive feelings and thoughts before you engage in conversation.

However, as Martin Seligman, the founder of positive psychology, points out, "Merely repeating positive statements to yourself does not raise mood or achievement very much." Instead, he says, you have to embed optimism in your brain "through the power of 'non-negative' thinking."[20] This means that you will need to consciously identify, then root out, the negative beliefs that have been unconsciously stored away in long-term memory.

You can begin this process by asking yourself what evidence there is to support your negative belief or fear. Often you'll find that your doubts are based on an exaggerated view of the situation. If you take a moment to pull yourself into the present moment, these old negative voices will lose their power.

Over time you can transform a helpless and pessimistic outlook into a realistic and lasting optimism. Positivity won't eliminate periods of depression, anxiety, and self-doubt, but it will dramatically reduce the number of incidents.[21] And this will improve every dimension of your relationships with others.

Step 5: Reflect on Your Deepest Values

In the previous chapter, we explored the transformational power of knowing your inner values. To set the right tone for a conversation, two other values that we've briefly mentioned need to be consciously addressed: your innermost relational value (for yourself in general and specifically concerning the person you're about to engage), and your deepest communication value (likewise, both for yourself and for the conversation you're about to have). Together these three values will create the best possible scenario when it comes to dealing with problems and achieving desired goals.

Few people hold anger and violence as values, but research confirms that sociopaths and people with antisocial and deviant behavior place the highest value on material gain and instant gratification.[22] Sometimes money and pleasure are their only values. Obviously, such people make bad risks for relationships that demand trust, integrity, honesty, kindness, and fairness—values that are essential in business and love.

If our personal, relational, and business values are not aligned with those of the person we are involved with, trouble is unavoidable. This suggests that we should ask others about their inner values as soon as we possibly can. But there's a catch: sociopaths are very good at reading other people's minds, and they can tell you, with great accuracy, what you hope to hear. They can also mask the nonverbal cues of deceit, so they're very hard to spot.[23]

However, when people become angry they act a little crazy. Like the sociopath, they become emotionally unpredictable, which makes it difficult to have a constructive dialogue. How do you communicate compassionately with angry people, staying true to your own inner values? It's difficult but not impossible. You have to identify, and then speak to, their underlying suffering and pain. You have to look *beyond* the anger. When you do this, as highly empathetic people can do, it will become easier to generate a compassionate smile that will help to defuse the anger being expressed by the other person.[24]

Ideally, when anger erupts a time-out should be called. But sometimes you can't do this. In such situations, it may help to focus on this question: what do I value most about this person? Then speak to those qualities. If you feel like you're about to lose your own patience or temper, then consider extricating yourself from the interaction as quickly as possible. Let the person know that you'll be happy to reengage when everything calms down.

Even if you enter a conversation with calmness, the other person's negativity may have more power because the primitive parts of your will brain kick into defensive and aggressive survival mode. They'll suck you in, and your positivity will vanish. Then what? Research says that you can deliberately suppress these negative reactions and arbitrarily impose a series of positive thoughts—on yourself and on the other person. This technique has been proven to be more effective than most of the other strategies that are used in anger-management training.[25]

Remember that verbal interaction often presupposes a goal-directed intention by the speaker.[26] To make a conversation balanced and fair, both parties need to be clear and up front, about values, intentions, and goals. Sharing these will make the communication process more efficient.

Step 6: Access a Pleasant Memory

It's best to enter a conversation with an inviting expression that conveys kindness, compassion, and interest. But as we explained in the previous chapter, this facial expression cannot be faked. It can be elicited by tapping into a pleasant memory, particularly one that involves people you deeply love and respect. This memory softens the muscles around your eyes and evokes a gentle half smile on your face.

When another person sees this expression, it stimulates a feeling of trust in their brain. The recollection of pleasant memories will also release pleasure chemicals throughout your own body and brain, and this will take you into an even deeper state of relaxation. When you look directly into the other person's eyes as you maintain this loving memory, they will *want* to engage you in a dialogue. Their facial expression will resonate with yours, and this will deepen the sense of contentment and satisfaction in both of you. As researchers at Loyola University Chicago demonstrated, contentment gives rise to mutually benevolent engagements.[27]

Why not just keep your face relaxed? Well, it turns out that a very relaxed face looks somber, which is why old photographs from the 1800s looked so unhappy. Back then, it took several minutes for an image to become fixed on the photographic plate, so a state of deep relaxation was the best way to keep a person's face still. In the early 1900s, when shutter speeds were faster, photographers were capable of capturing fleeting expressions of contentment.

Now you are ready to engage another person in a meaningful conversation, and it only takes about four minutes of preparation: a minute to stretch, relax, and yawn; another thirty seconds to bring yourself into the present moment; a moment to observe your inner speech and suppress it so that you can enjoy a few seconds of silence; another minute to fill your mind with positivity as you focus on your deepest values and goal; and finally the recollection of a memory that fills you with pleasure and joy.

With a little bit of practice, you'll be able to enter that exquisite state of heightened awareness in less than a minute or two.

Step 7: Observe Nonverbal Cues

"Keep your eyes on the ball." It's an expression used in sports and often applied to business, but when it comes to interpersonal relationships, it's essential to keep your eyes on the individual you are conversing with in order to discern the many nonverbal messages we constantly send to others. However, this does not mean that you should gaze unceasingly at the other person—that could feel invasive—but if you maintain softness in your eyes, generated by a pleasant memory, the other person won't want to take their eyes off you!

Eye contact stimulates the social-network circuits in your brain.[28] It decreases the stress chemical cortisol, and it increases oxytocin, a neurochemical that enhances empathy, social cooperation, and positive communication.[29]

Most people can recognize the seven basic facial expressions—anger, fear, sadness, disgust, surprise, contempt, and happiness—even though they remain on a person's face for just a few seconds. But as Ekman explains, you need to stay completely focused, making sure that you aren't distracted by your inner thoughts.[30]

If a person wants to conceal a feeling—out of embarrassment, discomfort, or the desire to deceive—the true expression might only appear for a quarter of second. Reading micro-expressions is not essential for effective communication; it simply gives you an edge. Nor will your impressions necessarily be accurate. You'll have to look for additional clues, and then ask the person if you are correct. But there's a problem: when you do this, the other person can feel violated. It's very disturbing when you discover that someone can read your mind.

Micro-expressions can only tell you that a true emotion is hidden, but it won't tell you why. Nor will it tell you whether the person is consciously or unconsciously concealing it. To ferret out these important bits of information, you'll have to talk more deeply with your partner.

When you learn how to read micro-expressions, says Ekman, "it gives you an edge in business because it allows you to communicate more effectively with business partners." We suggest that you visit Ekman's website (www.paulekman.com) to see how well you can detect micro-expressions using the micro-expression training tool. Ekman is currently using his research on facial expressions to help people cultivate emotional balance.

Step 8: Express Appreciation

The first words you speak will set the tone for the entire conversation, and a single compliment may be all you need to enhance cooperation and trust. Yet few people begin their conversations on a positive note. In fact, we're more inclined to speak out when we are bothered by something, not realizing that complaints immediately create a defensive reaction in the listener. So we have to train ourselves to bring as many expressions of appreciation into the conversation as possible. Every appreciative comment is a powerful form of affirmation and can reduce the negative mood of the recipient.[31]

Of course the compliment must be genuine, extending beyond the mere formality of a polite comment. As the staff at the Mayo Clinic emphasizes, "Relationships need nurturing. Build up your emotional account with kind words and actions. Be careful and gracious with critique. Let people know that you appreciate what they do for you or even just that you're glad they're part of your life."[32]

Our suggestion is to begin each conversation with a compliment but make sure that you end it with another compliment that conveys a deep sense of appreciation for the person and the dialogue you just had. Research shows that people respond better to compliments received at the end of an interaction than those given at the beginning of a dialogue.[33]

To make sure your compliments and statements of appreciation are genuine, we suggest you ask yourself this question: what do I really value about this person? As you contemplate that question, write down everything that comes to mind, and then ask yourself which, of all those attributes, you respect the most. Keep your answer in mind as you talk, and listen for an opportunity to share it. If such a moment doesn't occur, consider sending the person a note. An unexpected note of appreciation will rarely be perceived as a ploy.

Whenever I, Mark, turned in a manuscript to Jeremy Tarcher, my former publisher and personal friend, he always complimented it before suggesting how to make it better. The compliments always felt so genuine that I would fully embrace his suggestions. One day I asked him, "Do you really mean it when you compliment my writing, or are you just saying it because it's what an anxious writer needs to hear?" His response startled me: "Mark, I really don't know!" The moral of this story: when you make a habit of showing constant appreciation, even if it begins as a courtesy or subtle manipulation, your own mind comes to believe it's true.

Step 9: Speak Warmly

We cannot overemphasize the importance of speaking warmly—of conveying your compassion and sensitivity—but little research has been conducted on this element of communication. We know that different tones are registered and responded to by different language centers in the brain, but we're only beginning to identify which kinds of sounds reflect specific emotions and feelings.

In 2003 researchers doubted that we could map the human voice the way Ekman did with the face,[34] but now they feel more confident that emotions can be ascertained from nonverbal sounds. These "affect vocalizations," as they are called, may even be superior to facial expressions when it comes to telegraphing anger, contempt, disgust, fear, sadness, and surprise. However, facial expressions seem be more accurate expressions of joy, pride, and embarrassment.[35] Today we can identify many of the characteristics of vocal sound that express emotions and correlate them with the speaker's facial expressions.[36]

By looking for discrepancies between the face and the voice, we can come closer to identifying a speaker's truthfulness, sincerity, and trustworthiness, but we still do not have a documented way to train people to recognize many of the basic emotions concealed in tone of voice.[37] However, we can take some clues from actors, who have often been used in the research mentioned above. When actors need to project a warm demeanor, they do it by recalling a compassionate dialogue from their past.

If you drop the pitch of your voice and talk more slowly, the listener will hear and respond with greater trust. This strategy was developed and tested in 2011 at the Department of Communication Sciences and Disorders at the University of Houston, and it has been used to help oncologists present bad news to patients in the most supportive way possible. When the doctors reduced their speaking rate and pitch, the listener perceived them "as more caring and sympathetic."[38]

Ted Kaptchuk, of the Harvard Medical School, also discovered that using a warm voice would double the healing power of a therapeutic treatment.[39] Kaptchuk actually uses many of the strategies of Compassionate Communication to improve the health of his patients, and these, he states, are the key elements of success: "A warm, friendly manner; active listening . . . ; empathy . . . ; twenty seconds of thoughtful silence . . . ; and communication of confidence and positive expectation."

We use our words to express our wounds, and we use our words to heal. Thus it makes great sense that we train our voices to speak warmly, with confidence, empathy, and hope. Organizational psychologists at the University of Amsterdam concur: A strong, harsh, or dominant voice may impel others to comply with our wishes, but it will generate resentment that leads to weaker performance. A warm supportive voice is the sign of transformational leadership and will generate more satisfaction, commitment, and cooperation between members of a team.[40]

The Power of Emotional Speech

Study of the neural circuits associated with emotional speech gives us helpful information about strategies for speaking more empathetically.[41] For example, if you want to express joy, your voice needs to become increasingly melodic, whereas sadness will be conveyed by a flat and monotonic voice. When we are angry, excited, or frightened, we raise the pitch and intensity of our voices, and there's a lot of variability in both the speed and the tone.

However, if the vocal emotion is incongruent with the words you are using, it will create confusion for the listener.[42] You can test this by saying "I am angry" with a warm tone of voice and a sweet expression on your face. It creates a distinct pattern of neural dissonance. The same would be true if you heard "I love you" said in a loud, harsh tone of voice. At first the message would be confusing, but because the power of a negative word or sound trumps the power of a positive expression, the harshness would cause reactions of anger or fear in both you and the listener.[43]

Step 10: Speak Slowly

Slow speech rates increase a listener's ability to comprehend what you are saying, and this is true for both young and older adults.[44] Slower speaking will also deepen that person's respect for you,[45] and if you are speaking to someone with any form of language disability, it is essential to proceed slowly, articulating one word at a time.[46]

Interestingly, faster speakers are often viewed as more competent than slower speakers.[47] But we believe that this is a culturally learned behavior, and one that can easily be taken advantage of to mask a speaker's true intentions and inadequacies. Jeremy Dean, a researcher at University College London, suggests that we be particularly wary of the silver-tongued talker because "the fast pace is distracting and we may find it difficult to pick out the argument's flaws." He also adds that we should slow down when addressing our peers concerning matters of mutual agreement. [48]

Speaking slowly is not as natural as it may seem, and as children we automatically speak fast. But you can teach a child to slow down by speaking slowly yourself because they'll match the rate of your voice.[49] A slow voice has a calming effect on a person who is feeling anxious, whereas a loud, fast voice will stimulate excitement, anger, or fear.[50]

When we train people in Compassionate Communication, we ask participants to practice speaking extremely slowly so they can become aware of their speaking styles. The true power of speaking slowly is in the increased *consciousness* it brings to an otherwise habituated process.

Step 11: Speak Briefly

As you know by now, in Compassionate Communication we have a basic rule: whenever possible limit your speaking to thirty seconds or less. And if you need to communicate something essential to the listener, break your information into even smaller segments—a sentence or two—then wait for the person to acknowledge that they've understood you.

It's a hard concept to embrace. Why? The best reason we know of is that our busy minds have not been able to clearly formulate the essence of what we want to convey, so we babble on, externalizing the flow of information generated by our inner speech.

In centuries past, this problem was addressed by writing. If you really had something important to say, you wrote it in a letter or posted it in the community newspaper. Writing itself is a great way to consolidate one's thoughts, and so we recommend that you write down the major points of what you want to say, especially before an important meeting.

Although we've covered this point several times already, it bears repeating: our conscious minds can only retain a tiny bit of information, and for thirty seconds or less. Then it's booted out of working memory as a new set of information is uploaded. Our solution: honor the golden rule of consciousness and say only a sentence or two. Then pause and take a small deep breath, to relax. If the other person remains silent, say another sentence or two, and then pause again. This allows the other person to join in whenever they feel the need to respond or to ask for clarification. If you must speak for a longer period of time, forewarn the listener. This will encourage them to pay closer attention to you and to ignore their own intrusive inner speech.

Ideally, we suggest that you explain this rule of communication to your partner, and then invite them to experiment with you, each taking turns speaking a sentence or two for thirty seconds or less. If your partner agrees to this strategy, you'll find that you can accomplish an enormous amount in a short period of time, even if you don't use the other components of Compassionate Communication. This is the key strategy we teach to people who are involved in complex negotiations and conflict resolution, and it's especially effective when mediating volatile dialogues between opposing parties.

Step 12: Listen Deeply

To listen deeply and fully, you must train your mind to stay focused on the person who is speaking: their words, tone, gestures, facial cues—everything. It's a great gift to give to someone, since to be fully listened to and understood by others is the most commonly cited deep relationship or communication value.[51]

When the other person pauses—and hopefully they'll have enough self-awareness not to ramble on and on—you'll need to respond specifically to what they just said. If you shift the conversation to what you were previously saying, or to a different topic, it will interrupt the neurological "coherence" between the two of you, and the flow of your dialogue will be broken.[52]

When practicing Compassionate Communication, there's usually no need to interrupt. If the other person doesn't stop talking, they may be giving you an important clue. Perhaps their mind is preoccupied, or perhaps they are deeply caught up in their own feelings and thoughts. If this is the case, it's unlikely that they will be able to listen deeply to what you want to say.

But what if you have to convey something important, and your time is running out? Neurologically, this is a dilemma, because the listener will feel your interruption as an intrusion. There's no simple solution to this problem, which is why we encourage people to formally agree to speak briefly. If you must interrupt, you can apply the other strategies of Compassionate Communication. For example, you can quickly interject an apology and a compliment, using a warm, slow voice as you maintain a gentle gaze: "I'm sorry to interrupt since I do value what you are saying. But unfortunately I have a meeting I have to attend, and I want to make sure I'm able to tell you what I need to convey." For most people, this form of imposition will be met with appreciation.

It's also important to realize that most people are unaware that they are hoarding the conversation. They get caught up in their inner dialogues, and they are often impatient to speak lest they forget something important. In fact, research shows that most of us begin to speak *before* the other person has finished talking. Even doctors, who are trained to listen carefully for important medical information, tend to interrupt patients within twenty-three seconds, long before the patient's concerns have been stated![53]

Our advice: if you are engaged in an important discussion, and it becomes clear that the conversation is taking too long, you can suggest to the other person that you both take turns speaking only a sentence or two. You'll be surprised at how quickly an entire business plan, a medical treatment, or even a social event can be laid out.

If the other person keeps going on and on, and there's no need to interrupt, you can use this as an opportunity to study that person in detail. You can observe and at the same time watch how your own inner speech reacts. Allow yourself to flow with the words you hear and the facial expressions you see, and don't worry about what you may remember or forget. You'll actually be practicing a form of meditation that is neurologically enhancing and emotionally relaxing—a far cry from what we usually feel when we are bored by someone speaking.

Bad Listening

According to Lisa J. Downs, former president of the American Society for Training and Development, bad listening behaviors include daydreaming (thinking about unrelated topics when someone is speaking), debating (having an inner argument about what is being said), judging (letting negative views influence you), problem solving (yearning to give unasked for advice), pseudolistening (pretending to be a good listener), rehearsing (planning what you want to say next), stage hogging (redirecting the conversation to suit your own goals), ambushing (gathering information to use against the other person), selective listening (only responding to the parts of the conversation that interest you), defensive listening (taking everything personally), and avoidant listening (blocking out what you don't want to hear).

The Power of Intuition

There you have it: twelve steps and strategies that can transform any conversation into a remarkable event by fostering trust, empathy, and cooperation through the process we call neural resonance. But it will take practice to change the familiar patterns of dialogue that you are used to.

Effective communication demands a conscious, concerted effort, lest we slip back into old behaviors. So we ask you to practice these strategies at every chance you get and to share them with your family, friends, and colleagues. Discuss the twelve steps and decide which ones make sense for you. If you want to change them, by all means do, and if you find a strategy you believe is essential, please let us know. Compassionate Communication is a process and an "open source" experiment that hundreds of people have contributed to, and we expect that the process will continue to evolve.

This brings us to our final piece of advice, drawn from many years of research into the nature of human consciousness and the hidden powers of the mind: trust your intuition, and do what feels right for you.

Every person is unique, every interaction is unique, and every conversation is unique. Some strategies will work for some people at certain times, while other strategies will be called for with other people at other times. So we have to trust our intuition, which, from our perspective, contains a vast reservoir of insight that is rarely expressed in casual conversation.

Somewhere inside us—behind all the noise of everyday consciousness—there is a calm, observant self capable of making wise decisions. We can exercise this inner voice by practicing the twelve strategies of Compassionate Communication and following the advice of the inner wisdom of life.

CHAPTER 9

Compassionate Communication

Retraining Your Social Brain

In this chapter we're going to guide you through two specifically structured scripts designed to undermine old, habitual, and ineffective ways of communicating and replace them with a more effective strategy that will help to enhance social relationships, reduce misunderstandings and conflicts, and generate mutual cooperation and productivity for all concerned.

We've divided the script into two parts. The first part is structured so that you can do it alone, before you converse with someone, and the second part is designed so that you can practice Compassionate Communication with a partner.

After you've read through this chapter, we recommend that you practice this exercise with three different people and repeat it three times with each of them. That's nine practice sessions lasting twenty minutes each—a three-hour commitment—but it's enough to give you a thorough *experiential* sense of its value and to begin to build new neural networks in your brain.

The exercise is designed to alter the tempo and rhythm of speech while you remain in a relaxed state of heightened awareness. It integrates all the strategies we described in the previous chapter in a way that will allow you to bring them in real-life conversations with others. After you've practiced the training exercise below, you'll find that you can vary it to work with nearly everyone you meet.

Surprisingly, other people don't seem to notice that you're talking more slowly or more briefly or that you are observing their facial expressions more closely. Instead, if asked, they are likely to say that you appeared more focused and attentive. They sense that you have become a better listener, and you may perceive them as being more relaxed and interested in what you have to say. They will *feel* your presence, whether they are aware of it or not.

Preparing to Practice

The first ten-minute script will guide you through a relaxation and focusing process designed to bring you into the present moment, feeling positive about yourself and optimistic about engaging another person in a conversation. You can do this part alone, in preparation for talking with someone, or you can do it with a friend, family member, or colleague. The second ten-minute script is designed to be practiced with a partner who is willing to do this exercise with you. Practicing with a real person has its advantages, but it's not always practical.

Imaginary dialogues can have great value, especially when you need to address a difficult person or discuss an uncomfortable conflict with a family member or friend. By rehearsing how a conversation *may* evolve, you can often predict the best way to communicate your needs. You can also predict, with a fair degree of accuracy, how the other person might react or respond, and this allows you to alter your strategy in ways that would encourage a win-win scenario when the two of you sit down to speak. Imaginary dialogues have also been shown to entrain the brain in ways that will improve performance when they're actually carried out.

Ideally, the best way to follow any exercise is to listen to it, and in the index we've provided you with information for obtaining a CD or downloadable mp3 version of the Compassionate Communication script. When you read an exercise, you have to engage more neural circuits in your brain: you need to hold the book, recognize the words with your eyes, translate them into inner speech, and then carry out the step-by-step instructions. The solution: just give yourself more time as you read. Read each sentence out loud, very slowly. This will make the experience more meaningful and intense, and it will shift your consciousness into a more observant and attentive state.

Alternatively, you can make a recording of this script, as you practice it, and play it back. With today's computer and cell phone technology, it's easy to do. Just make sure that you speak as slowly as you possibly can. You can also take turns with a partner reading the script to each other. Or consider forming a small group, with several friends, and have one person read the script while others follow. This has the advantage of listening to how different people react and respond, and the positive experiences expressed by others will add support for practicing Compassionate Communication with others.

The Compassionate Communication Script

Let's begin. Find a comfortable chair and a quiet place to sit, where you won't be distracted by people or phone calls. If you have a bell or chime, keep it handy, and ring it when you see the asterisks in the script. It serves as an auditory reminder to slow down, relax, and pause. You can also go to http://www.mindfulnessdc.org/bell/index.html and click to hear the sound of a bell whenever you see an asterisk in the script.

The bell serves several functions: it entrains you go more deeply into a state of silent relaxation, and if you are practicing with another person, you can both use the bell as a reminder that the other person is talking for too long.

When reading the script below, you'll see asterisks to remind you to take an extra long pause before you go on to the next sentence. Use that pause to observe your feelings, sensations, and thoughts—noting them, and then letting them fade away into silence. With practice, you'll soon begin to integrate these conscious pauses into your everyday conversations, and the pause itself will enhance your neurological awareness, relaxation, and attentiveness.

Say each sentence just loud enough that you can hear your own voice, speaking warmly as you pronounce each word slowly:

> ***STEP 1:*** Take a few deep breaths, slowly breathing in and out through your nose.* Now relax all the muscles in your face. Let your forehead relax and let the muscles around your eyes relax. Take another deep breath and relax your jaw.* Now relax all the muscles in your neck.* Take a really deep breath and relax your shoulders. Lift your shoulders up to your ear, and then let them drop. Once more: pull your shoulders up high, hold them up, and then push them down toward the ground.* Breathe in deeply and feel how your chest moves in and out. Take another slow breath as you relax all the muscles in your arms and your hands. Feel the weight of this book in your hands, then close your eyes and use the next few moments to notice how your upper body feels.*
>
> ***STEP 2:*** Take a slow, deep breath and relax all the muscles in your back. Feel all the tension melting away, beginning with the top of your spine and slowly moving down to your hips.*

Take several deep breaths and feel how your stomach rises and falls.* Now turn your attention to your legs. Move them from side to side and feel yourself melting into the seat of your chair.* Now bring your attention to your feet. Tighten them, then relax them and shake them out.* Now turn your attention to your entire body. Tighten up all your muscles, hold them tense to the count of five, then let everything relax.* Once more: tighten up the muscles of your shoulders, arms, and legs. Hold them tight . . . tighter . . . and now relax.* Take a few slow breaths and mentally scan your body, looking for any excess tension in your muscles, your arms, your chest, your neck, and your face. Take another deep breath and let all your tension melt away.* Notice how your body feels and notice the calmness in your mind.

STEP 3: Now breathe in, and as you breathe out, make a soft sighing sound. Do this again, and then yawn. Yawn again, even if you don't feel like it, because it will make you even more relaxed and alert. Keep yawning for the next minute and notice how each one begins to feel more and more real.* Take another deep breath and feel how relaxed you are becoming. Slowly stretch your arms and legs, and slowly twist your torso from side to side. Do this for another minute in silence.* Pay attention to the sounds in the room; then bring your attention back to your breath and listen to the sounds of each inhalation and exhalation.*

STEP 4: Visualize someone you deeply love or recall a special moment from the past that fills you with pleasure and joy, something or someone that makes you feel like smiling.* Feel how the muscles of your face begin to change and notice how the muscles around your eyes are softening. Take a few deep breaths, then stretch again as you enjoy this feeling for the next few moments in silence.*

STEP 5: Ask yourself: what is my deepest, innermost value?* Notice whatever word comes to mind, then return your attention to your breathing.* Now ask yourself this: how important is it to me to be listened to and understood? Notice the thoughts and feelings that come to mind, then let them go.* Take a deep breath and relax some more.* Now ask: what tone of voice will best communicate what I want to say? Notice your thoughts, and let them go.* Shift your attention to your breathing and feel your whole body relax some more.* Now ask yourself: what is the best way to listen when someone else is speaking? Once more, notice the thoughts and feelings that come to mind.* Take a deep breath, stretch your body, and bring yourself back into the present moment.*

STEP 6: Visualize the person you plan to talk to. Think about a quality that you really like about that person and notice how it makes you feel.* Deepen your breathing as you hold a compassionate thought in your mind. Take a deep breath and continue to relax into your body.* Now visualize yourself talking to this person and paying them a compliment. Tell them what you appreciate about them and imagine that they have a warm gentle smile on their face.* Now imagine that this person is paying you a compliment. What do you hear them say?* What do you imagine they appreciate about you?* Notice how these imaginary comments make you feel.*

STEP 7: Next imagine that a compassionate dialogue begins to spontaneously unfold with this person. Visualize the two of you sitting there, taking turns speaking slowly and briefly, and fantasize about what is said.* One person speaks a sentence or two slowly, then stops and relaxes, bringing themselves back into the present moment. Then the other person speaks a few slow sentences and pauses, taking a deep breath and relaxing as much as they possibly can.* Keep this fantasy dialogue going for the next minute or two.*

STEP 8: Bring yourself back into the present moment as you let the fantasy conversation fade away. Take a few deep breaths and relax all the muscles of your face and body, and notice how your overall consciousness has changed.*

Entering a Conversation with Others

You are now ready to enter a real conversation. We recommend that you practice the script above as often as possible over the next few weeks before you engage in a dialogue with someone else. Also spend a few moments reviewing the twelve strategies of Compassionate Communication so you can bring them into every conversation. People have found it useful to write these words on an index card and to post it in various places as a reminder: on the computer or refrigerator, next to the phone, etc.

1. Relax
2. Stay present
3. Cultivate inner silence
4. Increase positivity
5. Reflect on your deepest values
6. Access a pleasant memory
7. Observe nonverbal cues

8. Express appreciation
9. Speak warmly
10. Speak slowly
11. Speak briefly
12. Listen deeply

When you walk into the room, slow down your gait; this too will serve as a reminder to remain consciously alert and relaxed. Greet the person with optimism and make certain that your words reflect a positive tone about the dialogue that is about to take place. As you wait for the other person to respond, make sure you stay relaxed as you observe the other person's face and body language. Listen to the tone of their voice and see if you can identify their emotional state. This information will let you know if the person can be responsive to what you want to say.

Practicing with a Partner

After you have experimented with the "imaginary conversation" script above ask a friend or family member or colleague to practice the exercise with you, along with the series of additional steps described below. When you first practice this technique with another person take two chairs, place them close together, and sit down facing each other. We suggest that you both read the steps out loud. You should be the first reader, since you have experienced the process and will be more comfortable and familiar with the exercise. This, by the way, is an excellent training device for learning how to engage another person in meaningful dialogue.

You will need to speak slowly so that the other person can become fully immersed in the experience. You can monitor the ideal speed by occasionally asking your partner if they want you to speak slower or faster. Some people feel uncomfortable with very slow speech but for most people the slower the better. You'll probably be surprised at the rate they feel most comfortable with, but since everyone is different, you'll need to tailor your speech to the needs of the listener. We recommend you do this when conversing with anyone.

Have your partner close their eyes if they feel comfortable doing so, because it helps most people to reach a deeper state of relaxed attention. Say one sentence slowly, then pause long enough for your partner to process the information as they explore their feelings and sensations. Tell your partner to lift a finger when they are ready for you to move on to the next sentence. Then, when you engage in the dialogue part of the exercise, either of you can lift your finger (or ring a bell) to remind the other person to slow down or talk more briefly.

In this first practice session don't talk about any difficult issues that either of you currently have. These can be addressed after you've practiced enough times that you both feel confident that you can remain in a relaxed and compassionate state when problems and conflicts are discussed. You can talk about your inner feelings and concerns, or about issues concerning other people, or about the positive aspects of your partner.

If you're not sure what to say or not to say, ask yourself the following question: can the other person hear and respond in a positive, compassionate way to what I want to say? If not, then the likelihood of having a conversation that leads to a positive outcome is limited. This is a rule that we suggest you apply, whenever possible, to all conversations you have.

Begin by guiding your partner through the first five steps. Ask them to keep their eyes closed. When you pause at each asterisk, shift your attention and study your partner's face. See if you can discern the subtle expressions that convey relaxation, pleasure, or frustration. If you sense any discomfort, ask them to take additional deep breaths and yawn, stretching and moving the parts of their body that you think might be tense.

As you continue to read, try lowering the tone of your voice and speaking a little softer. With each step, slow down a little more. When you have finished reading steps one through five, skip steps six through eight (they're designed for solo practice) and begin to read the script below.

> ***STEP 9:*** Visualize the person sitting across from you and smiling.⋆ Stay aware of your breathing as you hold a compassionate image or loving thought in your mind and deepen the relaxation of your body.⋆ Think about something you really appreciate about this person and notice how that makes you feel.⋆ Now think about another quality that you admire or respect. Again notice how this thought makes you feel.⋆ Affirm to yourself that the conversation you are about to have will be filled with compassion and respect.⋆ Now imagine such a conversation taking place. In your mind you see each person taking turns, speaking only one sentence at a time, slowly and briefly, then pausing. When one of you finishes speaking, you both take a deep breath and relax, pausing for about three seconds before the other person speaks.⋆ Imagine hearing the voice of your partner and no matter what that person says you feel yourself becoming more and more relaxed, and as you listen, your defenses continue to melt away.⋆

> ***STEP 10:*** Continue to visualize the conversation and imagine that a sense of trust and empathy is beginning to build. Let the conversation take any direction it wants. Don't control it; let it spontaneously evolve. Don't try to make a

point and don't worry if the subject changes. Just trust what your intuition brings up, staying as relaxed as you can.* There is no need to rush; all you have to do is talk slowly . . . then pause . . . then breathe . . . then listen . . . then relax . . . then talk again, with your voice filled with kindness and love.*

STEP 11: Now ask yourself this question: what's the most important value that I want to bring into the conversation I'm about to have?* Notice the thoughts and feelings that come to mind, then let them go, bringing yourself back into the present moment.* Now recall a memory that brings you a sense of pleasure, happiness, or joy.* Feel a warm smile begin to form on your face and feel the muscles softening around your eyes.*

STEP 12: In a moment you will open your eyes and gaze gently at your partner.* Begin the conversation by saying something appreciative or complimentary and listen to the compliment given by the other person.* Don't judge what they say. Just stay focused on your inner sense of well-being.* Let a spontaneous conversation emerge, speaking only one sentence, then stop and relax. Listen deeply as your partner speaks, and when they pause, take another deep breath and relax.* Then say whatever comes to your mind, but before your speak, ask yourself this question: will my words be met with kindness and appreciation?* If not—if you think they will make the other person defensive—take another deep breath and allow a different thought to come to mind.* Continue the dialogue for at least five minutes: talking, pausing, and listening as you stay as relaxed as you can.* Then close with a compliment and a statement of appreciation.

Put down your script and begin to talk with your partner. The conversation should flow quite naturally and easily, and since you'll each be speaking only one sentence, you'll be surprised how quickly it moves into areas that you probably would never have thought about discussing. Sometimes the conversation shifts to deep feelings and concerns about work or an incident that happened in the past. Sometimes a salient memory comes to mind. Or perhaps the conversation might focus on the values that matter most to you.

At other times nothing important comes to mind. If that happens, continue to stay relaxed and allow yourself to become comfortable with the "nonconversation." One of the purposes of this exercise is to interrupt the belief that we have to say something important or meaningful all the time. If we train ourselves to remain neutral, calmness itself will have a beneficial effect on the conversation. In essence you'll be learning how to be in the present moment with another person, without all the distracting thoughts that pull us away from our deeper nonverbal connection to each other.

Even remaining in silence as you gaze at each other, observing your own inner speech as you stay relaxed, is a powerful experience that has numerous psychological and neurological benefits.

The next time you practice Compassionate Communication you'll discover that the conversation will again take surprising turns. You can also begin to experiment with speaking longer—say two or three sentences at a time. When you do this, the goal is to make sure that the other person is fully engaged in listening to you. If not, it's a clear sign to pause and let the other person respond.

If one person has a tendency to talk longer than thirty seconds, you have several choices. You can simply observe to see if the other person realizes that they've engaged in a monologue, or you can watch how your inner speech reacts. This training exercise is a learning experience, and each time you practice, you'll intuitionally recognize layers of communication that you never noticed before.

If the person continues to speak too long, you can bring this issue into the Compassionate Communication dialogue. Tell them how you feel but tell them briefly, with a warm tone of voice. Many people have deeply embedded unconscious habits, and you may have to gently pull them into the Compassionate Communication model many times before their habits begin to change.

Going Deeper

If both parties want to, you can choose to take the conversation to a more intimate level. We suggest that you wait for about five to seven minutes before doing this, but we also suggest that you talk for no more than another five minutes. Close with a compliment and an appreciative comment. Then share your experience with other. What did you learn? What value do you feel the exercise has? What can you take away from it that you can incorporate into your daily conversations?

Now you can decide if you feel ready to discuss more difficult issues in your relationship. Often one person will feel ready to jump in but the other person will hesitate. Our advice: don't push it. Take more time and practice more rounds of Compassionate Communication. In those rounds talk about the concerns you may have, because this is the ideal state in which to establish the necessary ground rules for working through conflicts and problems. Use Compassionate Communication to create a game plan, and discuss how to handle the conversation if one person gets too upset.

The beauty of this exercise is that it allows you to cooperatively create strategies that are embedded with feelings of mutual trust and respect. If these qualities aren't present, then the communication process breaks down.

Usually the conversations that emerge are pleasant and constructive, but sometimes they bring up unexpected anxiety. For example, some couples begin to worry about what the other person might say. If you are conscious of this, then we suggest that you use the exercise to share those feelings, following the rule of not saying anything you think would be upsetting to the other person. It may take five, ten, or even twenty rounds of practice, but eventually progress will be made. If not, then a therapist (or, in the case of a business conflict, an executive coach) can probably assist. Continue to practice the strategies of Compassionate Communication, especially when a third party is involved.

Sometimes a person's inner anxiety or irritability will make them say something that interrupts your flow. If this happens, we suggest that you don't bring it up in the Compassionate Communication format until you've become proficient at it. Table the issue and use the "imaginary fight" exercise described in the following chapter.

Here's our suggestion for couples who want to make a commitment to this methodology. Decide together how often you want to practice each week. Ideally this would be five to seven times. When the other person isn't available, you can do steps one through eight in your imagination, and then rehearse what you want to say the next time the two of you talk.

Keeping a diary of your experience enhances the process, but don't use your diary to vent. Research has shown that this is quite a destructive act and that it will deepen your inner conflicts and anxiety.[1] In-

Compassionate Communication with Strangers?

For most people, the thought of sharing too much intimacy with people they barely know seems unwise. But research shows that speaking intimately to someone you don't know actually lowers stress and improves cardiovascular health! This was tested at the University of Southern Mississippi, with college students who were arbitrarily paired with one another.[2]

stead use your diary to construct positive solutions and track the positive progress you make.

Pay It Forward

We hope you'll experiment with this exercise with as many people as possible. First try it with your friends, then your family members, and don't be surprised if you find that your kids can follow the directions better than you! That's what we've found with our own children, as we'll illustrate in a later chapter.

Then try this exercise with a colleague at work. In fact, Compassionate Communication has proven to be very popular in the business community, to the point where many of our academic colleagues are writing articles describing ways to integrate the components into business management, educational training, and executive sales. Brevity, clarity, trust, and cooperation are essential for a company's financial success, and any strategy that can stimulate these qualities are readily incorporated into company policy.

Compassionate Communication is easy to teach, and in most sections of the community conducting workshops in it requires little formal training. In fact, the CD and mp3 we created work as a self-contained training module, and you can use them to guide an entire group of people. Even a single training session appears to be enough to enhance teamwork and improve conflict resolution in many social and work-related arenas.

One of the benefits of introducing Compassionate Communication in a group or classroom setting is the feedback that is generated. When you discuss what the optimal forms of communication *should* be, everyone comes away with new strategies that can be used to handle future problems and conflicts with greater effectiveness and resolve.

In the educational arena, it turns out that we rarely talking about talking. But it's essential when it comes to the people we're closest to, and so we recommend that you try the following experiment with your spouse, your kids, and with your closest friends and colleagues. Ask this question: what would you like me to change about the way I communicate with you that would improve our interaction?

In our experience this is a question that delights the person who is asked it. It gives them an opportunity to suggest ways to improve your interpersonal relationship. After all, don't we all want to know how to deepen our friendships and love? Yet when we ask roomfuls of people how often they've even considered asking such a question, hardly any hands are raised. Instead we hear a lot of nervous laughing, a sign that many already suspect what the other person might say. This suggests that we somehow know, deep down inside, that we are not talking and listening as well as we might.

Make this a topic during a formal Compassionate Communication exercise. Ask your partner what you can do to improve the process. We have found that both individuals will bend over backward to comply. Then, at your next compassionate dialogue, ask for feedback about how well you have done.

Once you become accustomed to the training exercise and the twelve strategies of Compassionate Communication, you can loosen up the rules. After all, there are times when it's really a delight to just sit back and chitchat. But if you do this with a little added awareness, even a superficial dialogue can bring deeper satisfaction and joy.

Finally, we would love to hear from you. Tell us about your experiences: the benefits you've gained, the problems you encountered, and any unique variations or new strategies that you have found useful. We'll try them out and test them, and share them with our growing community online. Our aim, as always, is to help people bring a little more peace into their lives and to bring that peacefulness into their conversations with others. In that way we hope, with your help, to bring a little more peace into the world.

PART 3

The Application

Practicing Effective Communication with Others

CHAPTER 10

Compassionate Communication with Loved Ones

In this chapter we'll share with you how different people have used Compassionate Communication to initiate dating, deepen intimacy, build empathy with strangers, and handle conflicts in their personal lives.

Every brain processes language in a different way, and this results in a communication style that is unique to each individual. Thus every conversation has the potential to unfold in a creative and original way. Normally, we try to standardize our language and the way we use our words, but as far as the research is concerned, this is nearly impossible to do. Different people continue to apply different meanings to the same words, and everyone uses a different style of vocalization. Some people like to speak as little as possible; others like to chatter away. Some prefer to talk superficially; others like to delve into the most personally revealing issues. These differences are what cause us to misunderstand one another, and the misunderstandings can lead to hurt, anger, and disappointment.

Compassionate Communication levels the playing field with the primary assumption that every conversation—indeed every sentence we speak—is filled with verbal and nonverbal messages that we frequently overlook. By bringing our awareness into the present moment, we can approach a conversation with an open perspective, the explicit intention to not impose our opinions or judgments on each other, and an honest desire to better understand the other person.

Compassionate Communication is not just about the dialogue. It's also about the *space* two people create during a conversation. You are practicing how to be *with* another person, in conversation, and in silence. When you honor this shared space, the relationship can take on a numinous quality that brings with it a heightened sense of awareness and aliveness.

A First Date

The best time to apply the twelve core principles of Compassionate Communication is when we first meet someone. They help us to suspend our inner speech and the natural anxiety that occurs when we meet a stranger or someone we don't know well. And they encourage us to feel open and relaxed enough to avoid putting the other person on guard. This is a moment when we need to be particularly observant so we can see if trust is even a possibility. Here's how a young man used Compassionate Communication when he began dating after his marriage collapsed.

It was a difficult time for George. He was separated and living alone, and he was about to go out on a date. He hadn't felt this vulnerable in years. A thousand voices were arguing in his head: "Am I still attractive? Will I like her? And if I do, will she like me or reject me? What should I say? What should I do? Oh God! I feel like a teenager in a middle-aged body." The voices went on and on.

He cringed at the idea of playing the dating game, but the notion of being alone for yet another weekend was just too much to bear. The woman he was about to meet was also going through a divorce, and they had been introduced through a local dating service.

George drove to the community park, where he and Marcy had decided to meet. He stepped out of the car and braced himself for the initial exchange. The inner voices were raging, and his heart was pounding in his chest. From head to toe, adrenaline filled him with fear.

In the past, dating had never been a problem because George used to be a player. He knew how to tease, and he knew how to seduce. He was good at it, but he always landed in bed with the wrong woman—someone who was as afraid of real intimacy as he was. He married a trophy girl, but the intimacy was never genuine, and so they fought a lot.

George didn't want to repeat the past, so he decided to do use the Compassionate Communication exercise, which he'd been introduced to at a seminar. He deliberately arrived at the park a half hour early and sat on a bench. Then he guided himself through a series of relaxation exercises.

Next he began to watch his thoughts. Each time a feeling of anxiety came up, he took a deep breath and relaxed some more. Then he told the thought to go away. "I don't believe you!" he said to his critical inner voice. It took about twenty minutes, but he finally calmed down. Then he envisioned staying relaxed as he imagined the person he was going to meet. More doubts popped up, but he pushed them out of his mind, replacing them with a positive affirmation. "I have no need to worry," he said to himself, "and I have no need to manipulate the date."

He stood up from the bench and walked slowly to the meeting place, and as he strolled along, he immersed himself in the beauty of the flowers and the sounds of the breeze blowing through the leaves of the trees. He was practicing staying in the present moment.

Then he caught sight of Marcy, sitting at their rendezvous spot. He kept his gaze on her, because this distracted him from the worrisome thoughts that began to race through his mind. He greeted Marcy, and they began a pleasant conversation, but George felt conflicted, afraid that he was using his old dating games. He wanted to tell Marcy what he was struggling with inside, but he knew that it could be risky, especially on a first date. Still, George decided to take a chance.

"What's the worst that can happen?" he reasoned to himself. "If she rejects me for sharing how I'm feeling, then I can move on and find someone who really wants to know who I am."

"Marcy," he began, "I'd like to try an experiment, if it's okay with you."

"What do you have in mind?" she cautiously replied.

"Would you be willing to talk for a few minutes using a technique I just learned in a workshop? The rules are simple: we'll speak very slowly and briefly, and we'll try to stay as relaxed as we can."

"Okay," said Marcy, but George noticed trepidation in her voice. This pleased him because in the past he never would have noticed such a subtle communication cue.

"I'll begin," said George. "I want to share with you how anxious I'm feeling. Normally I cover it up."

George noticed a look of surprise in Marcy's face, and he immediately felt more nervous. "This is a big mistake," he thought. Marcy then said something he didn't expect: "I know how you feel, but I'm pleased you told me that. Most of the men I've dated don't share what they really feel."

George felt a warm glow in his body and he smiled. For about twenty minutes, they talked about the difficulty of expressing emotions. Then George felt tears welling up in his eyes. "It's scary for me to say this to you, but I'm going to say it anyway: I feel like I've never been so open and honest with a woman before."

Marcy's eyes were also moist. She reached out and touched George's arm. "I know how you feel. My ex and I used to fight all the time, and I don't think we ever had this kind of intimacy, except on rare occasions. And here we're doing it on a first date!"

Marcy and George continued to talk slowly and intimately for the next three hours, and both agreed to use Compassionate Communication whenever a feeling of anxiety or anger cropped up.

And, yes, they eventually married.

The Last Date

Practicing Compassionate Communication does not guarantee that a relationship will blossom into love. In fact, when both people share how they really feel inside—when they explore each other's values in an atmosphere of respect—they may come to realize that they are not the best match. But if they continue to apply the principles of Compassionate Communication, they can end their relationship on a note of friendship, something that very few people manage to do. Often this may require the assistance of a therapist or coach.

Trudy had been married to Bob for five years, and they were like night and day. She loved children and animals. He didn't. She read spiritual self-help books and attended personal-development classes. He was critical of anything religious and preferred to unwind from his stressful work by watching TV. She liked to talk about everything, and he liked to complain. She tended to always look on the bright side of things, and he liked to preach doom and gloom. Even their political views were polar opposites.

Tensions built, he became depressed, and so she dragged him into couple's therapy. But during the first hour all Bob did was grumble to the therapist about the controlling nature of his wife.

The therapist, who was trained in Compassionate Communication, interrupted him with the following question: "Bob, tell me what your deepest value is?"

The question took Bob by surprise. "Honesty, I suppose."

"And your deepest relationship value?" asked the therapist.

Bob quickly responded, "Respect!"

The therapist said, "Bob, I know you're being honest when you tell Trudy how controlling she is, but let me ask you this: are you being respectful of her when you complain?"

With embarrassment, Bob mumbled, "Uh, I guess not."

The therapist then turned to Trudy. "What do you think of Bob's values: honesty and respect?"

"They're important, but I never feel that Bob treats me with respect," Trudy said with a strong note of hostility in her voice.

"Trudy," the therapist asked, "what are your deepest relationship values?"

"Kindness and intimacy," she immediately said.

"Bob, what do you think of Trudy's values?"

"I agree with those values," Bob replied.

"So we're all on board here," said the therapist. "I want the two of you to talk to each other about your concerns. You can be fully honest, but you both have to talk with deep kindness and respect. Do you think you can both do that?"

Bob and Trudy agreed.

"Great!" said the therapist. "But first I'm going to guide you through some exercises that will sidestep the anger you have been feeling for a while. Agreed?"

They did, and the therapist took them through the beginning steps of Compassionate Communication. When they both had serene expressions on their faces, the therapist asked them to talk about their deepest values. "I don't want either of you to bring up other issues today," he said. "We'll get to those later, when the two of you can speak to each other with kindness and respect." Then he explained the other components of Compassionate Communication.

"Now I want you to begin by telling each other what you love and respect about the other person," the therapist said.

Both Trudy and Bob struggled with this because it had been a long time since either had expressed kindness or gratitude to the other. Bob went first. "You're my best friend. I can tell you anything about myself and you really seem to listen."

Trudy was genuinely touched, but suspicious, so the therapist reminded her to take a deep breath and relax as she focused on personal memories of happiness.

"What I love about Bob is his honesty," Trudy said, looking at the therapist. "I can trust him in ways I could never trust my previous husband."

The therapist intervened. "Trudy, say that again but say it directly to Bob, as you gaze into each other's eyes." When she did, Bob smiled warmly, and an intimate dialogue quickly ensued. After several minutes Trudy reached over and touched Bob's hand—a good sign! After twenty minutes the therapist asked them how they felt about their relationship now, and they both replied that they had hope. They made a commitment to talk to each other every day for twenty minutes and to honor each other's values whenever they spoke.

Over the next eight weeks, Bob changed dramatically as he learned to interrupt his chronic negativity, and Trudy learned not to get caught up in her own critical inner voices. By observing her inner speech, she realized that she used criticism to distance herself from men, just as her mother had done. It was her way of running away and playing it safe. The relaxation exercises also helped Trudy to reduce her chronic anxiety.

At first the relationship seemed to flourish, but Trudy still felt a sense of emptiness inside. She began to think that her goals in life did not match her husband's, but she was afraid to share these thoughts with Bob. After all, he seemed so in love with her. Thus she was very surprised when Bob asked her if she might be happier living alone. By practicing the facial recognition strategies of Compassionate Communication, Bob had learned how to accurately intuit what Trudy was feeling inside.

Bob had also made a commitment to his therapist to spend ten minutes a day writing in a diary as he reflected on his deeper feelings, desires, and goals. He was asked to have imaginary dialogues with his wife but to interrupt any irritation he felt. Bob wasn't really an angry person; he had simply become so accustomed to it that he wasn't aware of how pervasive it was. All he had to do was to ask himself if he was really unhappy, and his response was always no. As he learned to reframe his negativity by making positive statements about how he actually felt, his entire mood changed and his self-esteem began to grow.

When you practice Compassionate Communication, you learn how to observe your partner without judgment. This allows partners to see who the other person really is, not what they wish or hope them to be. Bob and Trudy came to realize that they were always trying to please each other without taking care of their own needs first.

Sadly, their relationship reached a stalemate. They realized that their political and religious values were so different that they interfered with their social lives. They began to spend more time apart, pursuing their individual interests, and their romantic intimacy faded away. They were still friends but nothing more. Using the communication strategies they'd learned, they filed for divorce and separated their assets with fairness and mutual respect.

The good news: they both found new partners quickly, and the two couples became friends. Trudy's anxiety only returned on brief occasions, and Bob's gloominess disappeared. For the first time in years, he looked forward to going to work, and he took up a variety of recreational activities for the first time in his life.

What Makes Relationships Thrive?

Trust is one of the most important elements in a relationship because it can determine whether a relationship will succeed or fail. If you trust your partner, your relationship will thrive; if not, it won't.[1] Lack of trust leads to conflict, and conflict leads to what psychologists call "attachment anxiety."[2] In other words, quarrels and emotional arguments make it difficult for people to feel emotionally safe.

Low self-esteem and fear of rejection will also undermine relationship stability and trust.[3] In fact, the expression of any form of emotional anxiety and self-doubt acts as a signal to your loved ones or business associates that you are poorly handling interpersonal conflicts.[4] How can the other person tell? By reading the negative facial expressions that are generated by neural dissonance in your frontal lobes.[5]

Relationships thrive when people are immersed in an environment of positivity, mutual respect, cooperation, and trust. There's just no room for chronic negativity and self-doubt in business or in love.

Conflicts Damage Your Body and Your Brain

When marital conflicts occur, a common pattern is for one partner either to withdraw or immediately confront the other person. Both of these choices have a risk: an increase in cortisol,[6] a chemical known to increase stress and damage the brain and cardiovascular system. As researchers at the University of Utah found, "Marital conflict evoked greater increases in blood pressure, cardiac output, and cardiac sympathetic activation than did collaboration."[7]

Anger and hostility even interfere with the body's healing processes. Researchers at Ohio State College of Medicine brought forty-two married couples into a hospital and created small blister wounds on their arms. They measured the rate of healing and discovered that it took almost twice as long in couples who consistently communicated hostility toward each other.[8] Clearly, anger does not work well. However, positive communication between couples not only sped up the healing process of the wounds but also generated higher levels of oxytocin, the brain's love-and-bonding chemical.[9]

Our suggestion: the moment you feel tension building up inside of you, do everything in your power to physically and emotionally relax. If you can't, tell your partner that you need to take a time-out to unwind. This can be anywhere from ten minutes to a day or two, until the stress neurochemicals are eliminated from your body.

Having a mutual agreement to call time-outs provides a necessary safety valve when inner stress reaches the point that it spills into the conversation.

When conflicts are moderate or severe, a third party—a therapist, a friend, or in a work-related situation a neutral colleague—can be called in to mediate the conversation. The mediator points out subtle forms of negativity in the person's speech or behavior, then asks that individual to reframe their words in a positive, supportive way. The mediator can also monitor the rate of speech and use a bell to signal when a person needs to pause.

As researchers at the University of Rochester found, the more skilled we become at regulating our emotions while we speak, the more quickly we resolve our conflicts and with the least amount of stress.[10]

The Imaginary Argument

Conflicts are unavoidable, but as we become more adept at noticing nonverbal cues in a person's body, face, and vocal inflections, we can predict, with a great deal of accuracy, when a conflict is about to emerge. Furthermore, as we become more observant of our own inner speech and levels of tension, we can predict when we are most likely to get our buttons pushed.

This is where the power of imagination comes in, because if you have a fantasy conversation about a conflict that is starting to grow between you and another person, you can often identify what the best solution might be, before you engage in a dialogue.

Here's an effective way to develop this useful skill. Set up two chairs as though you were going to practice the Compassionate Communication exercise. Sit in one chair and face the empty one, imagining that the person with whom you are in conflict is sitting there. Then have an imaginary argument and watch where the conversation ends up. If you don't like the results, try another strategy. Communicate in a different way and play it out in your mind. Change the tone of your voice or try saying something complimentary to the imaginary person and see how they respond.

Now try getting really angry at the person and watch how they react in your imagination. Then ask yourself: if this were a real conversation, what would be the result? In this way you can often pinpoint conversational strategies that won't turn out well, and you can see how you might influence a more positive outcome by changing your style of talking.

It may take as long as an hour to find a strategy that will satisfy you, but we can guarantee that the approach you decide to use will be superior to what would have happened if you confronted the other person without this mental preparation. The "empty chair" exercise turns out to be more successful than other forms of interventions,[11] and the following "warm-up" exercise will help you to develop this skill.

> Imagine, for a moment, that someone—an old friend, lover, or business colleague (but not someone you currently interact with)—walks up to you and says something upsetting. Or think about a time in the distant past when someone you knew hurt your feelings or made you mad.
>
> Imagine that person walking up to you right now and hurting your feelings again. Use your memory to recall the feelings of anger, hurt, or pain. Keep focusing on the negative thoughts and feelings that come up and notice where in your body they affect you the most. Do they make your jaw tense up? Do you feel like making a fist or striking out or running away? Exaggerate the feelings and hold on to them for thirty seconds, but not longer.
>
> Now think about how you would normally react and notice how that makes you feel. Take a few deep breaths, relax your body, and let those thoughts and feelings float away.
>
> Now you ask yourself this question: when someone says something that upsets me, what is the best possible way to respond? Notice the thoughts that come to mind, then take another deep breath and relax.
>
> Again imagine someone saying something to you that would normally make you irritated, angry, and hurt. But instead of getting upset, imagine that you remain perfectly calm. Visualize the two of you standing there: the other person is yelling at you, but you are remaining completely relaxed and calm. No matter what the other person says, you continue to feel happy, joyful, and serene. Continue this visualization for as long as it takes you to really feel that sense of calmness.
>
> In your mind's eye, look at the angry person in front of you. Instead of focusing on the anger, try to see what is causing the other person to feel so upset. See if you can feel their hurt and pain, and then take a very deep breath and relax. Now speak out

loud to this imaginary person, and see if you can find the best words to make that person feel cared for, understood, and loved.

Notice how you feel, stretch a few times, and bring your attention back into the present moment.

The more you practice this imaginary exercise, the easier it will be to carry this relaxed and nondefensive state into a dialogue with someone else. The result, as research has shown: less interpersonal distress, fewer complaints, more mutual esteem, and more satisfying resolutions—at home and at work. And the effects have been shown last for at least a year.[12]

When you are ready to engage a loved one—or a colleague or friend—in a conflict-resolution dialogue, make certain you maintain a clear focus on the qualities you admire and respect about that person, in a manner that allows you remain true to your deepest values concerning relationships, communication, and love.

Asking Your Partner to Change

Compassionate Communication fosters self-awareness, but it doesn't give you the power to change the behavior of someone else. Only they can do that. You can let them know how their behavior makes you feel, but only if you communicate this in a way that does not assign blame. If you create the right atmosphere, using the strategies we've been describing, you can call for a special meeting to request a behavioral change. The other person may or may not comply, or even be able to change, but they will feel your kindness and respect.

The famous meditation teacher Thích Nhất Hạnh suggests trying the following strategy when you want to request a behavioral change.[13] Ask for an appointment to have a compassionate dialogue later in the week, and tell the person, with warmth and compassion, the issue you want to address. As he explains, this gives both of you time to prepare, to reflect deeply on the matter, and to be ready to enter the conversation with openness and trust:

> Suppose your partner says something unkind to you, and you feel hurt. If you reply right away, you risk making the situation worse. The best practice is to breathe in and out to calm yourself, and when you are calm enough, say, "Darling, what you just said hurt me. I would like to look deeply into it, and I would like you to look deeply into it, also." Then you can make an appointment . . . to look at it together. One person looking at the roots of your suffering is good, two people looking at it is better, and two people looking together is best . . . When you speak, you tell the deepest kind of truth, using loving speech, the kind of speech the other person can understand and accept. While listening, you know that your listening must be of a good quality to relieve the other person of his suffering.

Words of Love, Words of Hate

You can actually measure how stable a relationship is by counting the number of positive and negative emotional words that are used in everyday conversations. When researchers at the University of Texas analyzed the journals, e-mail, and text-messages of eighty-six young dating couples, those who included the most numbers of positive emotional words were more likely to still be dating three months later.[14] The message is clear: if you want your romantic relationships to last longer, send your partner as many heartfelt affirmations as you can. But they must be genuine, because the other person's brain is built to intuit lies.

Inflammatory words are particularly damaging to any relationship, whether it be at home or at work, and if you let your emotions take over during a marital conflict, they may literally break your heart. In a recent study reported in *Health Psychology*, forty-two couples were asked to talk about a topic that made them upset. Emotionally charged discussions caused the release of cytokines, proteins that are linked to cardiovascular disease, diabetes, arthritis, and various cancers. When people used words reflecting reason, understanding, and insight, the release of these stress chemicals went down.[15]

Our advice: choose your words carefully and be careful about ruminating on the conflicts in your marriage. This too releases stress chemicals that are damaging to your heart.[16]

Finally, some advice for everyone: avoid hostile words when you are around other emotionally volatile people; they can cause them to react with physical and emotional violence.[17]

Does Criticism Ever Work?

With few exceptions, the evidence says that criticism is not helpful, especially in spousal relationships. In fact, hardly anyone can tolerate criticism, especially from close relatives.[18] If there is already a degree of marital discord, criticisms will lead to greater unhappiness and strain,[19] and if you notice an increase in critical comments, it's a sure sign that the relationship is heading for trouble.[20]

It's important to recognize that different people can tolerate different levels of criticism. For example, some people seem to overreact to criticisms, and this may be a clue of an underlying depression.[21] On the other hand, some people don't realize that what they say would sound critical to the average individual; they just have thicker emotional skins.

Negative or destructive criticism means exactly what it says: you voice an objection or complaint that basically states that the person is wrong, mistaken, or bad without saying anything that might be helpful. Negative criticism strongly predicts marital discord and psychological symptoms, whereas constructive criticism is not perceived by the listener as being negative, critical, or dismissive.[22]

Constructive criticism requires that you don't voice a complaint or disapproval. Instead you offer a positive alternative approach that you feel may lead to a win-win interaction or solution. You can, for example, open the dialogue with a question like this: "I'm intrigued by your idea, but may I offer a different suggestion?" This type of question invites a positive response, and the person will usually say yes. When you address a problem in this manner, you show respect for the other person's opinion or behavior, even if it fundamentally differs from your own.

Criticizing other people rarely promotes cooperation and trust, but the real problem concerns the inner speech of everyday consciousness, because that is where the voices of self-criticism reside. The more self-critical you are, the more likely you'll become immersed in feelings of insecurity,[23] so it's important to recognize those voices and interrupt them in any way your can. Research shows that the most effective strategies involve the practice of self-love, self-appreciation, self-acceptance, and self-forgiveness, but you'll have to practice them on a daily basis if you want to extinguish the power of self-critical speech. Thus the first priority of Compassionate Communication is to teach your own inner voices how to get along with one another.

Who Are Better Communicators, Women or Men?

Men and women process language differently, they have different-sized brains and different neurochemical balances, but none of these differences translate into vast differences in behavior, memory, cognition, or verbal skills.[24] Men and women think, feel, and speak in essentially the same way.[25] The differences we see are superficial, culturally conditioned, or shaped by childhood experiences and adult biases. In reality, every person, whether male or female, has a unique style of thinking and feeling because no two human brains are wired the same way.

As the Smithsonian Institution reports, the differences we think exist are massively exaggerated: "When it comes to most of what our brains do most of the time—perceive the world, direct attention, learn new skills, encode memories, communicate (no, women don't speak more than men do), judge other people's emotions (no, men aren't inept at this)—men and women have almost entirely overlapping and fully Earth-bound abilities."[26]

Chapter 11

Compassionate Communication in the Workplace

Communication in the workplace is crucial to individual success and to the overall success of a company, and it begins the moment two people lay eyes on each other. In business first impressions matter. A recent brain-scan study found that one can even tell by looking at a CEO's face if he or she is trustworthy, has strong leadership skills, and is financially successful in governing the corporation.[1]

Of course, looks can also be deceiving. It's easy to mistake charisma—the ability to exude confidence—for competence. Charismatic leaders often spout values-based philosophies, and this tends to stimulate similar values in the people under their guidance.[2] But if the leader does not practice what he preaches, the followers' sense of being deceived will destroy the leader's credibility and possibly the credibility of the company itself. Just watch the stock market when a corporate leader violates a moral standard or gets caught up in the mystique of power and greed. These issues reflect the great importance of having and maintaining a strong sense of values.

Your Innermost Values Will Transform Your Work

Peter F. Drucker is an internationally renowned teacher best recognized for his popular books on business management, leadership, and entrepreneurship. He developed one of the first executive MBA programs in the United States, at Claremont Graduate University, where he was the Clarke Professor of Social Sciences and Management. At the age of eighty-nine, in an article published in *Harvard Business Review*, he said that if you want build a life of excellence, ask yourself these questions: "What are my values?" "What are my strengths?" And "What can I contribute?" Concerning values, he has the following to say:

> To be able to manage yourself, you finally have to ask, What are my values? This is not a question of ethics . . .
>
> . . . Ethics is only a part of a value system—especially of an organization's value system.
>
> To work in an organization whose value system is unacceptable or incompatible with one's own condemns a person both to frustration and to nonperformance.[3]

However, Drucker adds, "There is sometimes a conflict between a person's values and his or her strengths." What, then, should you do? Drucker believes that if you are not making a genuine contribution to yourself and the world, you should quit that job and search for another: "Values, in other words, are and should be the ultimate test."

Another corporate sage and renowned author is Marshall Goldsmith, recognized as one of the fifteen most influential business thinkers in the world. He teaches executive education at Dartmouth's Tuck School of Business, was the associate dean of the College of Business Administration at Loyola Marymount University, Los Angeles, and has coached some of the world's leading CEOs.

Goldsmith places strong emphasis on corporate and personal values, but he feels that such terms are bantered about with too much superficiality. Words like "quality," "integrity," and "respect," sound inspirational, but if action is not taken to back them up, they remain empty. He says, "There is an implicit hope that when people—especially managers—hear great words, they will start to exhibit great behavior."[4]

But they don't. The solution: get honest feedback from employees and respond to it with respect. Compassionate Communication sometimes fails in the business world because leaders do not want to give up their authoritarian control. If you don't honor and respect an employee's values and unique contributions, you cannot bring together a team of people and have them communicate effectively to reach mutual cooperation and satisfaction. Goldsmith puts it bluntly:

> As leaders we usually preach values involving people and teamwork but sometimes excuse ourselves from their practice. Even more often organizations fail to hold leaders accountable for living these values. This inconsistency invites corporate cynicism, undermines credibility, and can sap organizations of their vitality. The failure to uphold espoused values in general (and

Build Self-Esteem in a Week

This exercise was created by the University of Michigan's Stephen M. Ross School of Business. Ask ten to twenty people you know and trust—friends, colleagues, family members, customers, etc.—to give you a short description of the ways you add value to their lives. Ask them why they appreciate you and compose a brief essay consolidating the information you receive. You'll be building a portrait of who you are at your best.

"people" values in particular) is one of the biggest frustrations in the workplace.[5]

If we don't consciously discuss the value of values within the corporate environment, as an explicit dimension of company policy, how will our behaviors improve?

Maintaining the Connection

Not only do we have to communicate our values to others and act on them, we must do so in a way that shows how much we truly value them. In other words, leaders are responsible for instilling optimism and confidence in others. And this can only happen if we mutually honor each other's inner needs.

For example, researchers in the Management Department at Drexel University recently conducted a hundred-year profile study of seventy-five CEOs of major league baseball teams.[6] Those who encouraged confidence and optimism had teams that won more games and attracted more fan attendance. These CEOs also showed more concern for others than for themselves. CEOs that showed signs of conceit, vanity, and egotism ran teams that won the fewest games and attracted the least numbers of fans. Once again, we see that kindness and positive support makes all the difference in the workplace. This is particularly true for the health-care professions[7] and within the educational systems.[8]

The capacity to deeply relate to others is a key to all forms of relational success—at work and at home. If you find yourself in the position of overseeing others—be they your employees or your children—remember this: leaders who give the least amount of positive guidance to their subordinates are less successful in achieving their organizations' goals, and the employees are unhappier with their work.[9] Indeed by not taking an active role in dialogue and teamwork building, they generate more interpersonal conflicts within their groups.[10]

Bringing Compassionate Communication into Business Schools

Values-based leadership has become a priority in the business world, which is why Compassionate Communication was embraced by the Executive MBA Program at Loyola Marymount University, Los Angeles. Its strategies for reducing stress are an added bonus for people who have full-time jobs and have chosen to return to school to deepen their organizational skills.

Chris Manning, a professor of finance, points to the need to use brevity, clarity, and compassion in every aspect of business, leadership, and teaching: "In the classroom, I have learned that it is essential to create as much rapport with my students as possible. When I first began teaching thirty years ago, I usually spoke too fast, attempting to cover as much material as possible within the limited class time. This resulted in students being overwhelmed by the workload—an additional stress in an already stressful and difficult university class. Grades suffered, particularly for the weaker students, and some students would even drop out of the course. This experience taught me that we, as teachers and business executives, need to do everything in our power to show students and young corporate leaders why taking time out of their busy schedule to reflect on their personal and business values will improve their management skills with others. Just taking a few minutes each day to relax and be present can make their companies more successful. And if they don't bring these qualities into their business conversations and negotiations, sales will be lost and employees will quit."

Compassionate Negotiation

Deborah Kolb at the Simmons College School of Management emphasizes the importance of showing deep and genuine appreciation when negotiating with others: "Appreciative moves alter the tone or atmosphere so that a more collaborative exchange is possible." This, she adds, helps to ensure that all bargainers establish a common trust, away from "unspoken power plays and into the light of true dialogue."[11]

And remember: the more you communicate in a warm, supportive, enthusiastic, and genuinely caring way, the more you will be perceived as a transformational leader.[12]

Mindfulness, Stress, and Productivity

Herb Benson, of Harvard University, is one of the leading researchers exploring the neural mechanisms related to mindfulness, relaxation, and stress. He is using his findings to teach people how to get the most out of their work without getting burned out.

Benson, as mentioned above, discovered that a person can use "inner value" language to reduce physical and emotional stress. His well documented "relaxation response" uses the repetition of a single word or phrase—something that is highly meaningful to that person—to generate healthy changes throughout the body and the brain.

He calls his new technique the "breakout principle," and it helps hardworking people control their stress levels in a way that improves productivity and creativity. Here are the basic components, as described in the *Harvard Business Review.*[13]

First push yourself as hard as you can while working on a specific problem or a goal. Immerse yourself fully in the experience but stay aware of your level of stress. The moment you feel yourself tiring, take a break and go do something that is entirely unrelated to your work. Go for a walk, pet a dog, or take a shower. When you do this, the brain quiets down but, paradoxically, activity increases in the areas associated with attention, space-time concepts, and decision making. This can lead to sudden creative insight.

With practice, says Benson, you'll achieve a "new-normal state" of enhanced awareness and productivity, but only if you integrate stress-reduction and mindfulness strategies into your daily life—exercises like the ones we have presented in this book.

Increasing Positivity at Work

Marcial Losada is the director of the Center for Advanced Research in Ann Arbor, Michigan. His groundbreaking research shows that in the business world the most successful teams are those in which individuals are the most positive when communicating with one another. When they didn't like something that came up in the conversation, a negative person would show disapproval or sarcasm either in words ("That's a dumb idea!") or their facial expressions. A positive person would show support, encouragement, and appreciation toward the others, even if he or she disagreed with their plan. He or she might say something like this: "I understand what you are thinking, but let me explain why I think there is a better way." To respond in this manner takes skill and foresight, which is why we recommend taking a few extra seconds to silently rehearse what you are going to say, especially before responding to something you don't like.

Losada studied sixty business teams, and he found that the groups that showed a five-to-one ratio of positive to negative expressions were the most successful in business. Those that fell below a three-to-one ratio were the least successful.[14] Furthermore, people with high positivity ratios form stronger connections and bonds with others. They are grateful, upbeat, and likeable, and they regularly express compassion toward others. Negative people are irritable, contemptuous, and basically unpleasant to be around. Other research has shown that people who work under the command of a highly positive leader tend to be happier with their jobs.[15]

According to Losada and his research colleague Barbara Fredrickson, one of the cofounders of positive psychology, fewer than 20 percent of American adults generate a five-to-one positivity ratio, where one experiences "an optimal range of human functioning, . . . goodness, generativity, growth, and resilience."[16]

Clearly this is a call for all of us to promote positive thinking and communication whenever possible. How high can your positivity ratio go before the benefits level out? Eleven to one.

Compassionate Communication in the Medical Profession

When you work in a fast-paced industry like sales, the strategies of Compassionate Communication can increase your ability to resonate and empathize with your customer. The same holds true for the healing professions, especially in a hospital setting, where I, Andy, spend most of my working career. In this people-centered environment, the day-to-day stress can be so enormous that the extra time it takes to speak slowly can feel like a counterproductive strategy.

I often have to run between the hospital, the classrooms where I teach, and the lunchroom—assuming I have time to eat. But this undermines interpersonal relationships. When you're rushed, you're thinking about what you need to do next, not about what the other person is saying. But if you don't give your staff your fullest attention, oversights can be made that will affect their patients' lives. We have to slow down, even though we don't feel we have the time, or somebody may die.

Poor communication skills run rampant in the medical community, and you can see this beginning in the interview process for medical school applicants just as I see it in job applicants for my research team. Often I get one of two different kinds of applicants: those who speak too much and those who barely speak at all. Since it's my job to hire people who quickly and deeply connect with others, I study their nonverbal communications closely.

The talkers jump right in, giving me a synopsis of their entire life story. Sometimes they even talk about the weather or complain about another aspect of the interview process. I don't interrupt. I eventually say, "Well, the interview is over." This kind of applicant never allows me to establish any kind of relationship, and it costs them an opportunity for employment.

The nontalkers are another matter. I ask them important questions like "How do think we could improve the health-care system at this hospital?" They respond, "It's pretty complicated." I sit there and wait for more, but nothing else is said. Or I ask applicants about a project they are working on that could have value for the research I'm doing. You'd think they'd be thrilled to talk about their work, but all this type of applicant says is, "It's really interesting." I feel like I'm pulling teeth just to get them through an interview for a position that demands intense interpersonal dialogues with patients and hospital staff!

In both cases the interviewee's nervousness is usually the primary problem. Anxiety arousal causes some people to accelerate their speech rate and other people to freeze up.[17] For this reason, Mark and I have developed Compassionate Communication programs to teach beginning health-care professionals how to loosen up, make appropriate eye contact (which many applicants fail to do), and be present enough to engage in a meaningful two-way dialogue.

Researchers at the University of Southern Mississippi use a strategy very similar to Compassionate Communication. They train graduate students to improve poor interviewing behavior by using what they call "pause-think-speak." When asked a question, they identify the key words in the question, then make eye contact as they initiate a focused response.[18]

We've also begun to develop a Compassionate Communication stress-reduction program for surgeons at a well-known Southern California hospital. Surgery is an extraordinarily intense profession, and the burnout rate is high, but stress will also affect anyone's ability to perform well at work. Since lives are at stake, anyone who deals with emergencies—firefighters, emergency-room nurses, even plumbers who must rush out in the middle of the night to save a house from flooding—needs to be extremely calm and focused. Here's the technique we teach to surgeons and caregivers before they walk into the operating room or talk to a patient in need. It's equally applicable for anyone in business who is about enter into intense negotiations. Even a person who is about to haggle with a salesman can use this technique to negotiate a better deal.

1. Before you walk into the operating room, meeting room, or salesroom stop outside the door.
2. Take sixty seconds to yawn, stretch, and relax every muscle in your body.
3. Take a mental inventory. If you feel anxiety, irritation, or are distracted by unrelated thoughts, repeat step two until you are physically and emotionally calm.
4. Focus on your immediate goal and ask yourself: what is the frame of mind I need to be in? Suppress any negativity or doubts and envision yourself functioning at your very best.
5. Rehearse your strategy in your mind (research shows that

this improves performance when you carry out the actual task).

6. Focus on the values that mean the most to you as they specifically relate to your job or goal.
7. Relax your body some more, take a deep breath, and walk *slowly* into the room with a gentle smile on your face.

Even if you only have a minute to spare, stop for thirty seconds to relax and visualize a successful outcome, and do your best to maintain

Change Your Words, Change Your Life

In the ventures I've engaged in—as a U.S. Army officer during the Vietnam War, a financial executive, a venture capitalist, and an entrepreneur—I have often needed to generate dedication from my troops, employees, partners, or customers. I've found that the Compassionate Communication leadership techniques, as described in this book, help me to pull them out of their natural insecurity and help them to focus their creativity on devising new strategies to achieve our mutual goals. Even if there's only a minute to spare, we can use that minute to ground ourselves in body and mind. Speaking slowly and carefully will open the hearts of those you work with, and it will build good will with others.

I wasn't always this way. Twenty-five years ago I talked too fast and didn't give my fullest attention to what other people said. And the stress that caused was overwhelming. So I made a choice and changed my lifestyle, and I teach these lessons to my students. When you artfully apply the principles of compassion and bring it into your dialogues with others—especially in stressful situations—you'll achieve a better outcome in less time.

Chris Manning, Ph.D.
Professor of Finance and Real Estate
Loyola Marymount University, Los Angeles

that positive outlook as you attend to the task at hand. If the activity you're about to engage in requires dialogue, slow down your speech a little bit—just enough so that you can reflect on what you are going to say before you speak. This will promote focused, accurate, and brief communication that will have the greatest effect on those who hear your words.

To establish the best rapport with someone else, you should treat them with respect, pay attention to everything they say, and give them the best care and service you can.[19] In fact, it's your empathy—whether you are a caregiver, salesperson, or manager—that will create the greatest degree of mutual satisfaction.

Working in the Land of No

"When the principles of Compassionate Communication, as described in this book, are applied to business management, hiring, recruiting, or selling or integrated into any level of negotiation and work-related projects, superior results are invariably realized." This is the considered opinion of Stephen E. Roulac, a leading expert in strategic management, capital markets, and real estate investment. As an international business consultant, his clients include Apple Inc., Bank of America, Prudential, and the U.S. Department of Labor. He holds degrees from Stanford; University of California, Berkeley; and Harvard, and has held teaching positions at many universities. He has authored or edited twenty-two books and over four hundred articles. He is currently working with us to bring Compassionate Communication to a wider business audience. In reflecting on his values and career, he asked us if he could share this story with you. It exemplifies how each individual can personalize the components of Compassionate Communication and make it part of their work:

In my experience, having engaged in more than one million communications in virtually every aspect of business management, investment, and corporate decision making, one cannot afford to ignore the principles and strategies of Compassionate Communication. They need to be applied to one-to-one, real-time personal interactions, and they also need to be incorporated into the organizational message of the company. You cannot excel in the business world if you only honor one level of communication but disregard the other, for to do so would compromise the individuals as well as the corporate enterprise.

Some years ago I served as a senior management executive in a very large multinational firm. I regularly received written communications from the national office, which was abbreviated as "NO." Too often the messages themselves seemed to reflect the negativity of the acronym. Not only were they impersonal, but the content seldom reflected sensitivity. Rarely did they show any respect for, or even an acknowledgement of support to, the staff. The messages basically "barked" at you.

I suggested to the CEO of the firm that he could be more effective in realizing his objectives if he sent his messages from the position of "YES." I proposed that he relabel the national office as "Your Executive Services."

In advocating the change from NO to YES, I advised him that his audience would be more receptive to YES than NO. Even if his underlying message was not exactly what people might wish to hear, if he started from YES, he would be far better off than if he started from NO.

I felt that this YES position was particularly important because the main function of the corporate headquarters was to serve the firm's partners. The message of "Your Executive Services" would be both congruent and consistent with that priority. Since this particular CEO emphasized delivering outstanding client service, the YES framing would reinforce his strategic priority.

Sometimes in life—and especially in business—you learn as much by observing the effects of negativity as you do from positive modeling. This was one of those times, because this particular CEO of the NO school did not implement a positive YES communications positioning. In fact, he didn't even have the courtesy or capacity to acknowledge the suggestion. But then, what would you expect from the land of NO?

When you disregard the sound principles of Compassionate Communication, you compromise the strategic competence and integrity of the entire group. While this communication exchange—or should I say nonexchange—with the CEO was but one of many warning signals, I felt that it was profoundly symbolic and telling. I soon left the firm.

Our recommendation: take Compassionate Communication into your work and into the highest levels of management you can reach. Show them the research, experiment with your colleagues, and remember: it only takes one effective communicator—one compassionate leader or teacher—to cause a roomful of language-based brains to resonate to the quality of your speech.

And after you've introduced these strategies to your group, open the floor for debate. What you'll hear will move you in ways that you wouldn't expect, for as one CEO said, after practicing Compassionate Communication with the members of his board, "I never realized before what listening actually entailed."

CHAPTER 12

Compassionate Communication with Kids

We highly recommend that you experiment with the Compassionate Communication exercise with a child. Kids love it. For them, it's a game that puts them on equal footing with adults because the rules are the same for everyone. They especially like talking super slowly, and they're very good at limiting their conversations to thirty seconds.

When a colleague of ours, a math teacher at a local high school, did the exercise with his nine-year-old son, Nick, and his friends, he changed the rules and had a contest: each person wrote down an action-related sentence like "I'm going to ride my bicycle" or "I want a piece of pizza." Then they took turns trying to guess what the person was going to say as they spoke one word at a time, leaving ten seconds of silence between them.

Nick, for example, wrote down a sentence and folded the paper in half so that no one could see what he wrote. Then he said the first word: "My." The rest of the group yelled out sentences like "My stomach hurts." Obviously, no one could guess from just one word. Then Nick slowly said, "My . . . dad . . . talks." The responses became more focused: "My dad talks to my mom," etc. Nick then slowly said, "My . . . dad . . . talks . . . too." Immediately Nick's friend jumped in: "My dad talks too fast!" It was true, but not correct, so Nick's sister chimed in: "My dad talks too long!" Bingo!

Picking up on the cue, Nick's mom, a therapist, suggested that everyone play a round by starting out with a sentence that began with the name of someone in the room. As the game evolved, everyone learned something about how others perceived them. The game was a little risky, but with the direction of the parents, an atmosphere of fun was maintained, and the kids were able to express positive and negative thoughts about each other in a safe way.

The game also taught the children to pay close attention to the meaning of each word, and if they watched one another's faces and listened deeply to the tone of voice (the adults introduced these non-verbal communication concepts to the kids), they could be more accurate in predicting what another person would say. They were learning how to become more attentive of the subtleties of conversation and to fine tune their inner speech to stay focused on the meaning of other people's words.

A Mother-and-Daughter Dialogue

With the feedback we were getting, my wife, Stephanie, and I, Andy, became curious about how our eleven-year-old daughter, Amanda, would react. We talked to her about doing Compassionate Communication, and she seemed interested in trying it out, but, to be honest with you, I think she was more interested in being mentioned in this book! She loves to talk—with us, with her friends, and with other adults—so I thought she would be a natural.

We also had an important issue that we had been struggling with for years: when Amanda becomes hungry, her blood sugar drops. When this happens, she turns from her usual sweet self into a very unhappy person. The solution is simple: eat a snack or some food, but when she's in this grumpy state, it's very hard to get her to eat anything. She'll fight us in every way possible: "I don't feel like it! I'm not hungry! You can't make me! Leave me alone!"

Once she has eaten, it only takes about sixty seconds for this sullen, unpleasant kid to turn into an alert and excited person who loves to jabber away about all the things on her mind. But she never seems to remember this when she's hungry.

We'd discussed this issue with her many times in the past, but we'd never been able to adequately solve the problem. We hoped that the Compassionate Communication technique would help by giving everyone an opportunity to discuss how best to handle this issue together, as a family.

My wife and I agreed that she and Amanda would follow the training instructions on the CD because it was easier than reading the written instructions. I would observe and take notes, writing down what everyone said.

Amanda and Stephanie sat down and began listening to the CD. Amanda picked up the general approach rather quickly and wanted to jump right into it. I was fascinated to see what would happen at the ten-minute mark, after the relaxation and imagination exercises were presented and actual dialogue would start.

Amanda began with a compliment to her mom: "You are the most generous person I know." Both Stephanie and I were surprised. We had never heard her say this before. Stephanie responded with her own compliment: "I'm really touched by your willingness to play this game. You're such a wonderful child, and I'm really excited to do this with you."

Before I describe to you the results of the conversation, I want to point out something I found very interesting. Amanda loved to keep within the thirty second time frame, and she would frequently remind Stephanie to make sure that she did not talk too long. Occasionally, Stephanie wanted to talk more but Amanda wouldn't let her. Normally, Amanda loves to talk on and on and on, and so we were both intrigued by her willingness to comply with this rule.

After the compliments were exchanged, Stephanie asked Amanda what she thought about the problem of her forgetting to eat and how it affected her mood. Usually, this makes Amanda very defensive. But this time she responded with great calmness. And she really seemed to grasp what Stephanie was saying.

Amanda was actually able to explain to us why she normally found these discussions about her eating so annoying. She said—speaking slowly and briefly and staying within the time limit—that we were not listening to what she was trying to tell us in those situations.

As the conversation evolved, Stephanie and Amanda came up with some potential solutions. Amanda said that she would try to be more aware of when she is hungry, and both Stephanie and I agreed to be more responsive and to listen more closely to how she is feeling.

Overall, it was a very positive experience, and Amanda really enjoyed it. At the end she remarked that she was now aware that she was really hungry. So we went downstairs to get something to eat right away. No hesitancies, and no convincing, so day one was a success.

It's been a year since we had this "formal" dialogue, and Amanda continues to be much better about eating regularly, using her moods as a sign of when she needs food. She's also much better at modifying her mood when she does get hungry. Stephanie and I seem to be managing it much better as well, and we all seem to be listening to each other with greater empathy and understanding.

Starting Young

Most research supports the idea that our brain is heavily influenced by the environment in which we are born. Although a child younger than four or five really cannot engage in abstract conversations, from five on, they can. We also know that between the ages of five and ten, the brain is the most metabolically active it will ever be. The child's brain is forming and reforming billions of connections, especially those that relate to language and communication.

Research also shows that the more our brain is stimulated, through loving interactions with others, the more our neuronal connections grow. So it stands to reason that if we can engage our children at these early stages with lots of compassionate conversations, they will develop better communication skills from the get-go. This, we know, will translate into successful students and adults.

For example, Betty Hart and Todd Risely at the University of Kansas recorded more than thirteen hundred hours of interaction between parents and children from many different racial and economic backgrounds. Their findings, published in the book *Meaningful Differences in the Everyday Experience of Young American Children*, showed a direct link between a child's academic performance in third grade and the number of words spoken in their home from birth to age three.[1] They found that hearing about three thousand words per hour—about thirty thousand words a day—resulted in children being more successful later in life. In homes where the parents were professionals, this number was common, but in homes of lower socioeconomic status there was much more variability, ranging between five hundred and three thousand words per hour.

This means that over a year some children will hear over eleven million words while others will hear three million words or fewer. Ultimately, it doesn't matter if you are born rich or poor: what makes the difference between success and failure in life, between happiness and unhappiness, say the authors, "is the amount of talk actually going on, moment by moment, between children and their caregivers." The good news is that children from families of lower economic status whose parents did speak close to thirty thousand words per day showed the same results as their wealthier peers.

It's not just the quantity of words we use but the quality as well. Younger children first develop a larger vocabulary of negative words and have less ability to formulate positive words, especially when it relates to emotional states and the ability to achieve specific goals.[2] Yet negative words literally strain a child's brain.[3] They raise anxiety, whereas positive words lower a child's anxiety,[4] and for children who are under a great deal of stress, negative words will interrupt memory performance. They simply won't be able to recall the information that will best help them accomplish their goals. However, when we teach our children to use more positive words, we are actually helping their brains to have more emotional control and increased attention span.[5] And when we teach them the language of success, they become more motivated and satisfied with their work.

Parents who use a lot of negativity at home also undermine the stability of family life. When researchers at the University of Utah compared different styles of conflict resolution between parents and children, they found that cooperative *planning* with family members was more successful than using adult power to "lay down the rules." Thus authoritarian parents solved fewer problems with their kids.[6]

Between siblings, those who had positive verbal relationships with each other had fewer conflicts and were more likely to find creative solutions for their problems.[7] If one sibling took the lead and moved away from negative competition by offering positive solutions, the conversation would shift to a win-win scenario.[8] Thus by bringing the principles of Compassionate Communication onto the playground, and into peer-group mediation and training programs, we can effectively undermine the destructive emotional tendencies that many adolescents feel.[9] That's why we've begun to teach Compassionate Communication to a growing number of peer-to-peer support groups at colleges throughout the country.

We also need to teach students the rules of emotional intelligence, interpersonal intelligence, and intrapersonal intelligence so they can more quickly understand their own feelings, as well as those of others. When we do, we strengthen the communication processes that weave empathy, reason, and cooperation together in a meaningful way for groups.[10]

Compassionate Parenting

This all points to how important it is for parents to teach siblings how to use their optimism, serenity, and positive words to resolve conflicts with each other, and to learn how to recognize that every person sees the world in a different way.[11]

When parents bring the principles of Compassionate Communication into their families, their children show less aggressive behavior and get along better with their siblings.[12] When parents are taught how to listen deeply, they improve the dynamics with disruptive children.[13] In Pennsylvania a team of university researchers introduced families to a program that showed them how to "intentionally bring moment-to-moment awareness to the parent-child relationship" by "bringing compassion and nonjudgmental acceptance to their parenting interactions." They were taught to "pay close attention and listen carefully to their children," to "become more aware of their own emotional states and the emotional states of their youth," to "adopt an accepting, nonjudgmental attitude when interacting with their youth," to regulate their own emotional reactions during their interactions," and to "adopt a stance of empathy and compassion toward their children and themselves."[14]

They were also taught the same breathing, relaxation, and self-reflective exercises that we incorporate into Compassionate Communication, and they were shown how to focus their attention on deep listening. In teaching them how to avoid bringing anger or frustration into their conversations, they were given a simple phrase, "Stop, be calm, be present," as a reminder to control their negative emotions.

As a team of scholars from the University of Oxford, the University of Amsterdam, and Maastricht University have shown, compassionate parenting has the following benefits: it reduces stress, excessive worry, rumination, and negativity; it enhances attentiveness and promotes kindness and self-compassion; it improves marital satisfaction; and, perhaps most important, it breaks the cycle of passing on bad parental habits to the next generation.[15]

When young adolescents were asked to count their good moments and blessings, their sense of gratitude, optimism, and satisfaction increased.[16] They even felt more satisfied with going to school! However, if you choose to fill your diaries with details of your daily hassles, this will decrease your feelings of optimism and hope.[17]

The "Write" Way to Develop Positivity in Kids

As we mentioned before, just thinking about positive outcomes is not enough to build a solid foundation of optimism and self-esteem. Adults need to identify their unconscious negativity, reframe it, and repeatedly reaffirm it in positive words and actions.

For children and young adults, writing appears to be one of the most effective ways to achieve these important skills. High school students were asked to do the following task for ten days. Each night, before going to bed, they wrote down three things they did well that day. Then they stopped. At first not much improvement was seen, but with each passing month, for the next three months, the student's sense of happiness and well-being dramatically increased![18] And yes, it also has similar benefits for adults.[19] The author of these famous studies, Martin Seligman, who founded the field of positive psychology, added that the effects will not fade away, as is the case with placebos.

If just ten days of reflecting on what we do well can generate months of psychological improvement, imagine what would happen if you wrote down your accomplishments each day for a month? That's what we recommend you do, and to repeat this exercise anytime you feel frustrated in your work, your relationships, or your life.

Studies such as these also emphasize the power of the pen. In other words, it's not just your imagination that primes your brain for success. Writing deepens the impact by affecting different language centers in the brain thereby creating more permanent changes in how you think.

So if you want to transform a negative outlook on life, we suggest that you stimulate as many language centers in your brain as possible. Listen to positive words and messages. Read uplifting and encouraging novels. Think about the positive aspects and successes in your life and write them down. Then share your successes with others. Not only will it strengthen your own resolve, but it will also stimulate the listener's brain in positive ways.

But beware: the pen can be a double-edged sword. If you write down your negative feelings and thoughts, or write in your journal about stressful events, you'll tend to feel more emotionally distraught and report more symptoms of illness.[20] In fact, the more often you write about negative emotions, the more anxious and depressed you'll become.[21]

On the other hand, brief written commentaries about anxious

feelings can alleviate those symptoms temporarily, and as researchers at the University of Chicago discovered, "Simply writing about one's worries before a high-stakes exam can boost test scores."[22]

Here's another strategy children and adults can use to make positive changes. Keep a daily list of blessings and experiences for which you feel thankful. Research from around the world shows that this exercise will improve your mood and enhance your personal relationships.[23]

When 221 young adolescents were asked to keep a gratitude journal for three weeks, their sense of well-being, optimism, and satisfaction with life improved.[24] But when they kept lists of daily hassles, their moods and coping behaviors did not improve.[25] Children who feel the most gratitude toward others and about their lives exhibit greater satisfaction and optimism and have better relationships with their peers.[26]

When minority students wrote about themselves in positive ways, their sense of personal adequacy and integrity improved, along with their grades in school.[27] And if you write down your most important personal goals, as specifically as you possibly can, research shows that you'll be more likely to achieve them.[28] When we teach our children these strategies, the benefits will continue into adulthood, where they will be more successful than people who do not demonstrate a consistent positive attitude toward life.

The earlier we teach our kids how to practice Compassionate Communication, the easier our parental roles become. And because positive language and speech is contagious, we owe it to future generations to practice kindness whenever we interact with others.

Bringing Compassionate Conversations to the World

The practice of Compassionate Communication is an important step toward creating greater empathy and dialogue among all types of people in all sorts of circumstances. By fostering such dialogues, we have the opportunity to create a new and deeper understanding so that we can improve the society in which we live. Together we can create winning conversations where everyone benefits from the encounter.

When we change our words, we change our brain, and when we change our brain, we change the way we relate to others. The choice is ours: do we choose to spread negativity with our words, or do we choose to cultivate kindness, cooperation, and trust?

ACKNOWLEDGMENTS

Any book represents a collaboration of many voices, and this book is certainly no exception. First we would like to thank our students, patients, and workshop participants: without your willingness to share your stories, experiences, struggles, and inspirations this book would have never come into existence.

We would like to thank our colleagues, friends, and family members, all of whom contributed endless hours of time to slowly refine the strategies described within this book. Special thanks is extended to Neil Schuitevoerder, Ph.D., one of the original codevelopers of Compassionate Communication who continues to work with us to bring this strategy into the therapeutic community.

Our deepest appreciation also to Dorianne Cotter-Lockard, Ph.D., who coauthored an academic article on Compassionate Communication with us and who presented the paper at the annual conference of the American Psychological Association in 2010. We want to acknowledge the generous assistance of Chris Manning, Ph.D., and William Lindsey, Ph.D., for bringing Compassionate Communication into the Executive MBA Program at Loyola Marymount University, Los Angeles, and to thank John Baker, Ph.D., and Paul Mattson, Ph.D., at Moorpark College for bringing us into their classrooms to conduct our workshop research. To the ministers and congregations of the United Centers for Spiritual Living (with a special thank-you to Rev. Pam Geagan), Unity Church, the Unitarian Universalist Association of Congregations, and the many Christian and secular organizations with whom we've worked: warm blessings to all for your support.

Finally, we would like to express our profound gratitude to our agent, Jim Levine, and our editor, Caroline Sutton, for believing in this project and supporting us through every stage of this book. We also wish to thank all the wonderful people behind the scenes at Penguin and Hudson Street Press who helped to give birth to this book.

APPENDIX A

Compassionate Communication Training: CDs, Mp3s, Workbooks, Webinars, and Workshops

To aid you in the practice of Compassionate Communication, we have created a seventy-minute self-guided CD (and an mp3 downloadable file) to complement this book. It is designed to teach individuals, couples, and groups how to change their talking and listening behaviors in ways that facilitate mutual trust, empathy, and comprehension. In addition to the twenty-minute Compassionate Communication training module described in chapter 9, it will guide you through the inner values exercise described in chapter 7, a kindness and forgiveness meditation, and a series of stress-reduction and movement exercises designed to guide you into a deep state of relaxation.

In the first ten minutes of the training module, you'll be guided through the first six steps of Compassionate Communication, in which you engage in either an imaginary or real dialogue with another person. In the second ten minutes (designed to be played while you practice with a friend, colleague, family member, or in a group situation), a bell will ring every twenty-five seconds to remind you to slow down, stop speaking, and return to a relaxed state as you listen to the other person speak. The program also includes additional strategies for communicating effectively and resolving conflicts. This self-guided program can be used to train large groups of people in a business, church, or school setting.

A complementary program (available as a CD or mp3 download) is also available to help reduce stress and deepen your Compassionate Communication practice. It includes seven relaxation and mindfulness exercises that research has proven effective and that can be used by individuals or introduced to students in the public and private school systems. This stress reduction program is now part of the Executive MBA module that Mark teaches at Loyola Marymount University, Los Angeles.

To order with a credit card either of these programs in CD format, call Mark's office at (805) 987-7222. To order these programs as a downloadable mp3, go to www.MarkRobertWaldman.com, where you will find additional videos, programs, workbooks, and free materials relating to personal development, business coaching, and executive communication.

At the Mindful Living Foundation (www.MindfulLivingFoundation.org) you can view videos relating to neuroscience and listen to meditations and lectures created by leaders in the mindfulness community.

If you would like to have Mark speak at your group, contact him at MarkWaldman@sbcglobal.net and visit his website if you want to attend any of the workshops he offers throughout the year: www.MarkRobertWaldman.com.

APPENDIX B

Compassionate Communication Training Protocol for Couples

If you would like to try the formal training program we use in our research, and which therapists use to help couples develop stronger communication skills, here is a simplified version. Feel free to send us descriptions of your experiences, copies of your diaries, or any suggestions you may have. We're always looking for ways to improve our strategies, and the people using them are our most valuable source for inspiration and guidance. We also have a growing network of therapists and business coaches who can offer personal assistance via telephone or video conferencing. For further information, go to www.MarkRobertWaldman.com or call Mark's office at (805) 987-7222.

To begin this eight-week training program, it's best if you purchase a copy of the Compassionate Communication CD or downloadable mp3, but you can also improvise using the exercises in this book. Ask your partner or a friend to commit to practicing with you for at least three to five days a week. The first three sessions should not be focused on any specific issue; just let a spontaneous conversation emerge. Then pick a specific issue that the two of you would like to address. Start with a simple problem and when you feel confident with the process, begin to tackle the more difficult ones. However, if either of you feels anxious or irritable, take a break. When you feel ready, use the Compassionate Communication strategy to talk about your struggle, not about the problem itself. Explore possible ways to address your problem in a manner that both people feel is safe. If you still feel that you are at an impasse, consider having a Compassionate Communication coach guide you through the issue.

With your partner present, follow the instructions below:

1. Listen to track 3 on the Compassionate Communication CD (or do the inner values exercise described in chapter 7). Share your inner values with each other.
2. Now listen to track 7 on CD (or read the Compassionate Communication script in chapter 9), which will guide you through the dialogue process with your partner. A suggestion: consider a dialogue based on your inner values.
3. After completing the exercise, take a sheet of paper and spend no longer than five minutes writing down your subjective experience. How did you feel during the exercise? What did you find useful? What made you uncomfortable? What did you discover? Then have a five-minute discussion with your partner focusing only on the positive aspects of your experience.
4. Listen to the "Kindness and Forgiveness" track on the CD (or listen to Jack Kornfield's meditation on compassion that is posted at www.MindfulLivingFoundation.org) with your partner. When finished, talk about the positive aspects of your experience.
5. Beginning tomorrow, over the next five days listen to all the tracks on the CD and repeat steps one through three, above. If possible, do so with your partner, but if he or she is not available, listen to them alone and use your imagina-

tion to guide yourself through the dialogue exercise (track 7 or the instructions in chapter 9).

6. After each practice round, write several brief paragraphs in a journal. If you find yourself writing down a complaint or negative thought, limit the comment to one sentence and then write several positive ways that you can overcome or reframe the negative feeling or thought. Keep each journal entry brief, not longer than a couple of short paragraphs. Do not share this diary with your partner; it is for your own personal exploration.
7. Keep a daily log of the time spent doing the exercise, whom you did it with (alone, partner, etc.), the topic discussed or imagined, and the feelings it brought up.
8. Each day, write down three to five things you feel grateful for.
9. At the end of each day, write down three things you did well that day and briefly explain why.

At the end of eight weeks, we would love to hear from you. Send a letter or an e-mail with a description of your experience (and your diaries, if you feel comfortable) to markwaldman@sbcglobal.net. You don't need to include your name, just your gender and age. All information received will be kept confidential, and we will analyze the data in ways that will help to improve our strategies. Again, you have our deepest gratitude for taking time to participate in this transformational experiment.

APPENDIX C

Compassionate Communication Research Study

In 2010 we performed our own pilot study of 121 people engaging in a Compassionate Communication workshop. At the beginning of the workshop and again at its conclusion, participants were given a questionnaire called the Miller Social Intimacy Scale (a validated measure of a person's feelings of closeness and social empathy). Using basic statistical analysis, we were able to detect a significantly higher level of intimacy after participants had practiced the exercise for forty minutes. The scores were actually about 11 percent higher. In responses to two questions, we found a 20 percent improvement in social intimacy: "How close do you feel to [the person you are facing] right now?" and "How strong is your inclination to spend time alone with him/her?"

We also wondered if there were any differences among people of different ages. When comparing a younger community college (CC) group to the control group, the results showed that the CC group had a mean increase of only 6 percent. So this younger population appeared to respond less robustly in general; however, we did achieve a significant increase compared to the control group (the group of people who were tested before and after forty minutes of exercise). However, people over the age of thirty generally responded to a much greater degree, showing a mean increase of 16 percent.

From the point of view of race, very similar results were obtained for both whites and blacks. The black participants came from a church congregation in a lower-income community in downtown Los Angeles, and the white participants from churches in highly affluent suburbs. The community college students represented a mixed socio-economic background. From this limited data set, we inferred that socio-economic background had no definite influence on measures of social intimacy.

In terms of gender, both men and women did about the same. At baseline women scored slightly higher on their intimacy level, but this did not achieve statistical significance. The increases in the level of intimacy after the training program were similar for both men and women.

We added one additional question to the Miller questionnaire: "What is your secret desire?" Participants were asked the same question before and after they practiced Compassionate Communication, and when we conducted a content analysis to determine how often certain words were used, we found that after the exercise interest in financial goals dropped from 34 percent to 14 percent. More importantly, there was a 60 percent increase in a desire for peace, while desires for self-love and interpersonal love increased threefold. This strongly suggests that the Compassionate Communication program improves intimacy and may even redirect a person's goals toward more peaceful, loving, and personal fulfillment. This research study is currently in submission for peer-reviewed publication.

NOTES

Chapter 1: A New Way to Converse (pages 3–22)

1. Manning C, Lindsey W, Waldman M, Newberg A. Paper prepared for presentation at the 28th Annual Meeting of the American Real Estate Society. April 2012, St. Petersburg Beach, Florida.
2. Levinson S. C. *Presumptive Meanings*. MIT Press, 2000.
3. Sperber D, Wilson D. *Relevance: Communication and Cognition*, 2nd Ed. Blackwell Publishers, 2001.
4. "Causal impact of employee work perceptions on the bottom line of organizations." Harter J. K., Schmidt F. L., Asphund J. W., Killham E. A., Agrawal S. *Perspectives on Psychological Science*. 2010; 5(4):378–89.
5. "Creative innovation: Possible brain mechanisms." Heilman K. M., Nadeau S. E., Beversdorf D. O. *Neurocase*. 2003 Oct; 9(5):369–79.
6. "Cognition without control: When a little frontal lobe goes a long way." Thompson-Schill S. L., Ramscar M, Chrysikou E. G. *Current Directions in Psychological Science*. 2009; 18(5):259–63.
7. "Neurocognitive mechanisms underlying the experience of flow." Dietrich A. *Consciousness and Cognition*. 2004 Dec; 13(4):746–61. See also: Csikszentmihalyi M. *Flow: The Psychology of Optimal Experience*. Harper, 1991.
8. "Neural and behavioral substrates of mood and mood regulation." Davidson R. J., Lewis D. A., Alloy L. B., Amaral D. G., Bush G, Cohen J. D.,

Drevets W. C., Farah M. J., Kagan J, McClelland J. L., Nolen-Hoeksema S, Peterson B. S. *Biological Psychiatry*. 2002 Sep 15; 52(6):478–502.

9. "How anger poisons decision making." Lerner J. S., Shonk K. *Harvard Business Review*. 2010 Sep; 88(9):26.
10. "Functional projection: How fundamental social motives can bias interpersonal perception." Maner J. K., Kenrick D. T., Becker D. V., Robertson T. E., Hofer B, Neuberg S. L., Delton A. W., Butner J, Schaller M. *Journal of Personality and Social Psychology*. 2005 Jan; 88(1):63–78.
11. "Portrait of the angry decision maker: How appraisal tendencies shape anger's influence on cognition." Lerner J. S., Tiedens L. Z. *Journal of Behavioral Decision Making*. 2006; 19: 115–37.
12. Fredrickson B. *Positivity*. Three Rivers Press, 2009.
13. "Positive affect and the complex dynamics of human flourishing." Fredrickson B. L., Losada M. F. *American Psychologist*. 2005 Oct; 60(7):678–86.
14. "Is there a universal positivity bias in attributions? A meta-analytic review of individual, developmental, and cultural differences in the self-serving attributional bias." Mezulis A. H., Abramson L. Y., Hyde J. S., Hankin B. L. *Psychological Bulletin*. 2004 Sep; 130(5):711–47.
15. "Anterior cingulate activation is related to a positivity bias and emotional stability in successful aging." Brassen S, Gamer M, Büchel C. *Biological Psychiatry*. 2011 Jul 15; 70(2):131–37.

Chapter 2: The Power of Words (pages 23–38)

1. "Some assessments of the amygdala role in suprahypothalamic neuroendocrine regulation: A minireview." Talarovicova A, Krskova L, Kiss A. *Endocrine Regulations*. 2007 Nov; 41(4):155–62.
2. "Happiness and time perspective as potential mediators of quality of life and depression in adolescent cancer." Bitsko M. J., Stern M, Dillon R, Russell E. C., Laver J. *Pediatric Blood and Cancer*. 2008 Mar; 50(3):613–19.
3. "The role of repetitive negative thoughts in the vulnerability for emotional problems in non-clinical children." Broeren S, Muris P, Bouwmeester S, Van der Heijden K. B., Abee A. *Journal of Child and Family Studies*. 2011 Apr; 20(2):135–48.
4. "Protocol for a randomised controlled trial of a school based cognitive behaviour therapy (CBT) intervention to prevent depression in high risk adolescents (PROMISE)." Stallard P, Montgomery A. A., Araya R, Ander-

son R, Lewis G, Sayal K, Buck R, Millings A, Taylor J. A. *Trials*. 2010 Nov 29; 11:114.

5. "What is in a word? No versus yes differentially engage the lateral orbitofrontal cortex." Alia-Klein N, Goldstein R. Z., Tomasi D, Zhang L, Fagin-Jones S, Telang F, Wang G. J., Fowler J. S., Volkow N. D. *Emotion*. 2007 Aug; 7(3):649–59.
6. "Neural mechanisms of grief regulation." Freed P. J., Yanagihara T. K., Hirsch J, Mann J. J. *Biological Psychiatry*. 2009 Jul 1; 66(1):33–40. Epub 2009 Feb 27.
7. Wright, R. *The Moral Animal: Why We Are, the Way We Are: The New Science of Evolutionary Psychology*. Vintage, 1995.
8. "Erasing fear memories with extinction training." Quirk G. J., Paré D, Richardson R, Herry C, Monfils M. H., Schiller D, Vicentic A. *Journal of Neuroscience*. 2010 Nov 10; 30(45):14993–97.
9. "Generalized hypervigilance in fibromyalgia patients: An experimental analysis with the emotional Stroop paradigm." González J. L., Mercado F, Barjola P, Carretero I, López-López A, Bullones M. A., Fernández-Sánchez M, Alonso M. *Journal of Psychosomatic Research*. 2010 Sep; 69(3):279–87.
10. "Negative and positive suggestions in anaesthesia : Improved communication with anxious surgical patients." Hansen E, Bejenke C. *Anaesthesist*. 2010 Mar; 59(3):199–202, 204–6, 208–9.
11. "In search of the emotional self: An fMRI study using positive and negative emotional words." Fossati P, Hevenor S. J., Graham S. J., Grady C, Keightley M. L., Craik F, Mayberg H. *American Journal of Psychiatry*. 2003 Nov; 160(11):1938–45.
12. "Genomic counter-stress changes induced by the relaxation response." Dusek J. A., Otu H. H., Wohlhueter A. L., Bhasin M, Zerbini L. F., Joseph M. G., Benson H, Libermann T. A. *PLoS One*. 2008 Jul 2; 3(7):e2576.
13. "Neural correlates of abstract verb processing." Rodríguez-Ferreiro J, Gennari S.P., Davies R, Cuetos F. *Journal of Cognitve Neuroscience*. 2011 Jan; 23(1):106–18.
14. "Modulation of the semantic system by word imageability." Sabsevitz D. S., Medler D. A., Seidenberg M, Binder J. R. *NeuroImage*. 2005 Aug 1; 27(1):188–200.
15. "Neural evidence for faster and further automatic spreading activation in

schizophrenic thought disorder." Kreher D. A., Holcomb P. J., Goff D, Kuperberg G. R. *Schizophrenia Bulletin*. 2008 May; 34(3):473–82.

16. "Neural correlates of long-term intense romantic love." Acevedo B. P., Aron A, Fisher H. E., Brown L. L. *Social Cognitive and Affective Neuroscience*. 2011 Jan 5.
17. "May I have your attention, please: Electrocortical responses to positive and negative stimuli." Smith N. K., Cacioppo J. T., Larsen J. T., Chartrand T. L. *Neuropsychologia*. 2003; 41(2):171–83.
18. "On the incremental validity of irrational beliefs to predict subjective well-being while controlling for personality factors." Spörrle M, Strobel M, Tumasjan A. *Psicothema*. 2010 Nov; 22(4):543–48.
19. "The value of positive psychology for health psychology: Progress and pitfalls in examining the relation of positive phenomena to health." Aspinwall L. G., Tedeschi R. G. *Annals of Behavioral Medicine*. 2010 Feb; 39(1):4–15. "Positive psychology in clinical practice." Lee Duckworth A, Steen T. A., Seligman M. E. *Annual Review of Clinical Psychology*. 2005; 1:629–51.
20. "Positive psychology progress: empirical validation of interventions." Seligman M. E., Steen T. A., Park N, Peterson C. *American Psychologist*. 2005 Jul–Aug; 60(5):410–21.
21. "What is in a word? No versus yes differentially engage the lateral orbitofrontal cortex. "Alia-Klein N, Goldstein R. Z., Tomasi D, Zhang L, Fagin-Jones S, Telang F, Wang G. J., Fowler J. S., Volkow N. D. *Emotion*. 2007 Aug; 7(3):649–59.
22. "Happiness unpacked: Positive emotions increase life-satisfaction by building resilience." Cohn M. A., Fredrickson B. L., Brown S. L., Mikels J. A., Conway A. M. *Emotion*. 2009 Jun; 9(3):361–68.
23. "Detecting deceptive discussions in conference calls." Larcker D, Zakolyukina, A. Stanford Graduate School of Business Working Paper 2060: July 29, 2010.
24. "Affective habituation: Subliminal exposure to extreme stimuli decreases their extremity." Dijksterhuis A, Smith P. K. *Emotion*. 2002 Sep; 2(3):203–14.
25. "Genomic counter-stress changes induced by the relaxation response." Dusek J. A., Otu H. H., Wohlhueter A. L., Bhasin M, Zerbini L. F., Joseph M. G., Benson H, Libermann T. A. *PLoS One*. 2008 Jul 2; 3(7):e2576.

26. "Increased BDNF promoter methylation in the Wernicke area of suicide subjects." Keller S, Sarchiapone M, Zarrilli F, Videtic A, Ferraro A, Carli V, Sacchetti S, Lembo F, Angiolillo A, Jovanovic N, Pisanti F, Tomaiuolo R, Monticelli A, Balazic J, Roy A, Marusic A, Cocozza S, Fusco A, Bruni C. B., Castaldo G, Chiariotti L. *Archives of General Psychiatry*. 2010 Mar; 67(3):258–67.
27. "The effects of subliminal symbiotic stimulation on free-response and self-report mood." Weinberger J, Kelner S, McClelland D. *Journal of Nervous and Mental Diseases*. 1997 Oct; 185(10):599–605.
28. "Evaluative priming from subliminal emotional words: Insights from event-related potentials and individual differences related to anxiety." Gibbons H. *Consciousness and Cognition*. 2009 Jun; 18(2):383–400.
29. "Murder, she wrote: Enhanced sensitivity to negative word valence." Nasrallah M, Carmel D, Lavie N. *Emotion*. 2009 Oct; 9(5):609–18.
30. "Evidence of subliminally primed motivational orientations: The effects of unconscious motivational processes on the performance of a new motor task." Radel R, Sarrazin P, Pelletier L. *Journal of Sport and Exercise Psychology*. 2009 Oct; 31(5):657–74.
31. "When sex primes love: Subliminal sexual priming motivates relationship goal pursuit." Gillath O, Mikulincer M, Birnbaum G. E., Shaver P. R. *Personality and Social Psychology Bulletin*. 2008 Aug; 34(8):1057–69.
32. "The neural basis of love as a subliminal prime: An event-related functional magnetic resonance imaging study." Ortigue S, Bianchi-Demicheli F, Hamilton A. F., Grafton S. T. *Journal of Cognitive Neuroscience*. 2007 Jul; 19(7):1218–30.
33. "Predicting persuasion-induced behavior change from the brain." Falk E. B., Berkman E. T., Mann T, Harrison B, Lieberman M. D. *Journal of Neuroscience*. 2010 Jun 23; 30(25):8421–24. http://www.ncbi.nlm.nih.gov/pmc/articles/PMC3027351/?tool=pubmed.
34. "Neural activity during health messaging predicts reductions in smoking above and beyond self-report." Falk E. B., Berkman E. T., Whalen D, Lieberman M. D. *Health Psychology*. 2011 Jan 24.
35. "Grasping language—A short story on embodiment." Jirak D, Menz M. M., Buccino G, Borghi A. M., Binkofski F. *Consciousness and Cognition*. 2010 Sep; 19(3):711–20.

36. "Characterization of fear memory reconsolidation." Duvarci S, Nader K. *Journal of Neuroscience*. 2004 Oct 20; 24(42):9269–75.
37. "A prospective study of cognitive emotion regulation strategies and depressive symptoms in patients with essential hypertension." Xiao J, Yao S, Zhu X, Abela J. R., Chen X, Duan S, Zhao S. *Clinical and Experimental Hypertension*. 2010 Dec 19.
38. "Ethical principles and economic transformation—A Buddhist approach." Zsolnai L. *Issues in Business Ethics*. 2011; 33, part 4.
39. "Gross national happiness." Tideman S. G. *Issues in Business Ethics*. 2011; 33, part 3.
40. "Neuroeconomics and business psychology." Larsen T. *China USA Business Revue*. 2010 Aug; Vol 9.
41. "Prediction of all-cause mortality by the Minnesota Multiphasic Personality Inventory Optimism-Pessimism Scale scores: Study of a college sample during a 40-year follow-up period." Brummett B. H., Helms M. J., Dahlstrom W. G., Siegler I. C. *Mayo Clinic Proceedings*. 2006 Dec; 81(12):1541–44.

Chapter 3: The Many Languages of the Brain (pages 39–52)

1. Galaburda A. M., Kosslyn S. M., Eds. *The Languages of the Brain*. Harvard University Press, 2002.
2. "Invariant visual representation by single neurons in the human brain." Quiroga R. Q., Reddy L, Kreiman G, Koch C, Fried I. *Nature*. 2005 Jun 23; 435(7045):1102–07.
3. Papousek H, Jürgens U. *Nonverbal Vocal Communication: Comparative and Developmental Approaches* (Studies in Emotion and Social Interaction). Cambridge University Press, 1992.
4. Deacon C. *The Symbolic Species: The Co-Evolution of Language and the Brain*. Norton, 1998.
5. "Where are the human speech and voice regions, and do other animals have anything like them?" Petkov C. I., Logothetis N. K., Obleser J. *Neuroscientist*. 2009 Oct; 15(5):419–29.
6. "Twitter evolution: Converging mechanisms in birdsong and human speech." Bolhuis J. J., Okanoya K, Scharff C. *Nature Reviews Neuroscience*. 2010 Nov; 11(11):747–59.

7. "A meta-analytic review of gender variations in adults' language use: Talkativeness, affiliative speech, and assertive speech." Leaper C, Ayres M. M. *Personality and Social Psychology Review*. 2007 Nov; 11(4):328–63.
8. "When language meets action: The neural integration of gesture and speech." Willems R. M., Ozyürek A, Hagoort P. *Cerebral Cortex*. 2007 Oct; 17(10):2322–33.
9. "Gestures orchestrate brain networks for language understanding." Skipper J. I., Goldin-Meadow S, Nusbaum H. C., Small S. L. *Current Biology*. 2009 Apr 28; 19(8):661–67.
10. Ekman P. *Emotions Revealed*. Holt, 2007.
11. "Intention processing in communication: A common brain network for language and gestures." Enrici I, Adenzato M, Cappa S, Bara B. G., Tettamanti M. *Journal of Cognitive Neuroscience*. 2010 Oct 18.
12. "Memory effects of speech and gesture binding: Cortical and hippocampal activation in relation to subsequent memory performance." Straube B, Green A, Weis S, Chatterjee A, Kircher T. *Journal of Cognitive Neuroscience*. 2009 Apr; 21(4):821–36. "Neural interaction of speech and gesture: Differential activations of metaphoric co-verbal gestures." Kircher T, Straube B, Leube D, Weis S, Sachs O, Willmes K, Konrad K, Green A. *Neuropsychologia*. 2009 Jan; 47(1):169–79.
13. "Good and bad in the hands of politicians: Spontaneous gestures during positive and negative speech." Casasanto D, Jasmin K. *PLoS One*. 2010 Jul 28; 5(7):e11805.
14. "Embodiment of abstract concepts: Good and bad in right- and left-handers." Casasanto D. *Journal of Experimental Psychology: General*. 2009 Aug; 138(3):351–67.
15. "How our hands help us learn." Goldin-Meadow S, Wagner S. M. *Trends in Cognitive Sciences*. 2005 May; 9(5):234–41.
16. "When the hands speak." Gentilucci M, Dalla Volta R, Gianelli C. *Journal of Physiology* (Paris). 2008 Jan–May; 102(1-3):21–30. Epub 2008 Mar 18.
17. "Spoken language and arm gestures are controlled by the same motor control system." Gentilucci M, Dalla Volta R. *Quarterly Journal of Experimental Psychology* (Colchester). 2008 Jun; 61(6):944–57.
18. "Mirror neurons and the evolution of language." Corballis M. C. *Brain and Language*. 2010 Jan; 112(1):25–35. Epub 2009 Apr 1.

19. "How symbolic gestures and words interact with each other." Barbieri F, Buonocore A, Volta R. D., Gentilucci M. *Brain and Language*. 2009 Jul; 110(1):1–11.
20. Ekman P. *Emotions Revealed*. Holt, 2007.
21. "Where is the love? The social aspects of mimicry." Van Baaren R, Janssen L, Chartrand T. L., Dijksterhuis A. *Philosophical Transactions of the Royal Society of London: Series B, Biological Sciences*. 2009 Aug 27; 364(1528):2381–89. "Conversation and coordinative structures." Shockley K., Richardson D. C., Dale, R. *Topics in Cognitive Science*. 2009; 1(2):305–319.
22. "Gesture gives a hand to language and learning: Perspectives from cognitive neuroscience." Kelly S, Manning S. M., Rodak S. *Language and Linguistics Compass* 2 (2008):10.1111/j.1749–818.
23. "Communicating emotion: Linking affective prosody and word meaning." Nygaard L. C., Queen J. S. *Journal of Experimental Psychology: Human Percepttion and Performance*. 2008 Aug; 34(4):1017–30.
24. "Domestic dogs use contextual information and tone of voice when following a human pointing gesture." Scheider L, Grassmann S, Kaminski J, Tomasello M. *PLoS One*. 2011; 6(7):e21676. Epub 2011 Jul 13.
25. "Reduction in tonal discriminations predicts receptive emotion processing deficits in schizophrenia and schizoaffective disorder." Kantrowitz J. T., Leitman D. I., Lehrfeld J. M., Laukka P, Juslin P. N., Butler P. D., Silipo G, Javitt D. C. *Schizophrenia Bulletin*. 2011 Jul 1.
26. "Mark my words: Tone of voice changes affective word representations in memory." Schirmer A. *PLoS One*. 2010 Feb 15; 5(2):e9080.
27. "Tasty non-words and neighbours: The cognitive roots of lexical-gustatory synaesthesia." Simner J, Haywood S. L. *Cognition*. 2009 Feb; 110(2):171–81.
28. "A sweet sound? Food names reveal implicit associations between taste and pitch." Crisinel A. S., Spence C. *Perception*. 2010; 39(3):417–25.
29. "As bitter as a trombone: Synesthetic correspondences in nonsynesthetes between tastes/flavors and musical notes." Crisinel A. S., Spence C. *Attention, Perception, and Psychophysics*. 2010 Oct; 72(7):1994–2002.
30. "Neural substrates of attentive listening assessed with a novel auditory Stroop task." Christensen T. A., Lockwood J. L., Almryde K. R., Plante E. *Frontiers in Human Neuroscience*. 2011 Jan 4; 4:236. "Insula, a 'mysterious'

island in our brain—Minireview." Palkovits M. *Orv Hetil.* 2010 Nov 21; 151(47):1924–29.

31. "Functional connectivity of the insula in the resting brain." Cauda F, D'Agata F, Sacco K, Duca S, Geminiani G,Vercelli A. *NeuroImage.* 2011 Mar 1; 55(1):8–23.
32. "Long-term meditators self-induce high-amplitude gamma synchrony during mental practice." Lutz A, Greischar L. L., Rawlings N. B., Ricard M, Davidson R. J. *Proceedings of the National Academy of Sciences of the United States of America.* 2004 Nov 16; 101(46):16369–73.
33. "Meditation experience is associated with increased cortical thickness." Lazar S. W., Kerr C. E., Wasserman R. H., Gray J. R., Greve D. N., Treadway M. T., McGarvey M, Quinn B. T., Dusek J. A., Benson H, Rauch S. L., Moore C. I., Fischl B. *Neuroreport.* 2005 Nov 28; 16(17):1893–97.
34. "Cerebral blood flow changes associated with different meditation practices and perceived depth of meditation." Wang D. J., Rao H, Korczykowski M, Wintering N, Pluta J, Khalsa D. S., Newberg A. B. *Psychiatry Research.* 2011 Jan 30; 191(1):60–67.
35. "Sentence comprehension in autism: Thinking in pictures with decreased functional connectivity. "Kana R. K., Keller T. A., Cherkassky V. L., Minshew N. J., Just M. A. *Brain.* 2006 Sep; 129(Pt 9):2484–93.
36. "Imagery in sentence comprehension: An fMRI study." Just M. A., Newman S. D., Keller T. A., McEleney A, Carpenter P. A. *NeuroImage.* 2004 Jan; 21(1):112–24.

Chapter 4: The Language of Consciousness (pages 53–75)

1. "Neural correlates of consciousness reconsidered." Neisser J. *Consciousness and Cognition.* 2011 Apr 12. "Intraoperative awareness: From neurobiology to clinical practice." Mashour G. A., Orser B. A., Avidan M. S. *Anesthesiology.* 2011 Apr 1:1218–33.
2. "Consciousness and neuroscience." Crick F, Koch C. *Cerebral Cortex.* 1998 Mar; 8(2):97–107.
3. "Exploring consciousness through the study of bees." Koch C. *Scientific American.* January 14, 2009.
4. "The global workspace (GW) theory of consciousness and epilepsy." Bar-

tolomei F, Naccache L. *Cognitive and Behavioral Neurology*. 2011 Jan 1; 24(1):67–74.

5. "Cortical midline structures and the self." Northoff G, Bermpohl F. *Trends in Cognitive Sciences*. 2004; 8(3):102–7. "Reflective self-awareness and conscious states: PET evidence for a common midline parietofrontal core." Kjaer T. W., Nowak M, Lou H. C. *NeuroImage*. 2002; 17(2):1080–86.
6. "Individual working memory capacity is uniquely correlated with feature-based attention when combined with spatial attention." Bengson J. J., Mangun G. R. *Attention, Perception, and Psychophysics*. 2011 Jan; 73(1):86–102.
7. "Penrose-Hameroff orchestrated objective-reduction proposal for human consciousness is not biologically feasible." McKemmish L. K., Reimers J. R., McKenzie R. H., Mark A. E., Hush N. S. *Physical Review E: Statistical, Nonlinear, and Soft Matter Physics*. 2009 Aug; 80(2 Pt 1):021912.
8. Watch the excellent video documenting the Princeton University PEAR program exploring how the mind influences objects at a distance: http://www.princeton.edu/~pear/.
9. "Compassionate intention as a therapeutic intervention by partners of cancer patients: Effects of distant intention on the patients' autonomic nervous system." Radin D, Stone J, Levine E, Eskandarnejad S, Schlitz M, Kozak L, Mandel D, Hayssen G. *Explore* (NY). 2008 Jul–Aug; 4(4):235–43.
10. "Double-blind test of the effects of distant intention on water crystal formation." Radin D, Hayssen G, Emoto M, Kizu T. *Explore* (NY). 2006 Sep–Oct; 2(5):408–11.
11. "The emergence of human consciousness: From fetal to neonatal life." Lagercrantz H, Changeux J. P. *Pediatric Research*. 2009 Mar; 65(3):255–60.
12. "The birth of consciousness." Lagercrantz H. *Early Human Development*. 2009 Oct; 85(10 Suppl):S57–58.
13. "Functional pathophysiology of consciousness." Jellinger K. A. *Neuropsychiatry*. 2009; 23(2):115–33.
14. "Epilepsy, consciousness and neurostimulation." Bagary M. *Behavioral Neurology*. 2011 Jan 1; 24(1):75–81.
15. "The development of color categories in two languages: A longitudinal study." Roberson D, Davidoff J, Davies I. R., Shapiro L. R. *Journal of Experimental Psychology: General*. 2004 Dec; 133(4):554–71.

16. "Color vision: Color categories vary with language after all." Roberson D, Hanley J. R. *Current Biology*. 2007 Aug 7; 17(15):R605–7. "Color categories: Evidence for the cultural relativity hypothesis." Roberson D, Davidoff J, Davies I. R., Shapiro L. R. *Cognitive Psychology*. 2005 Jun; 50(4):378–411.
17. "Language and perceptual categorization." Davidoff J. *Trends in Cognitive Sciences* 5 2001:383–87.
18. "Neurophysiological mechanisms and consciousness." Creutzfeldt O. D. *Ciba Foundation Symposium*. 1979; (69):217–33.
19. "Self-recognition, theory-of-mind, and self-awareness: What side are you on?" Morin A. *Laterality*. 2010 Nov 3:1–17.
20. "Opposing effects of attention and consciousness on afterimages." Van Boxtel J. J., Tsuchiya N, Koch C. *Proceedings of the National Academy of Sciences of the United States of America*. 2010 May 11; 107(19):8883–88. "The relationship between awareness and attention: Evidence from ERP responses. "Koivisto M, Kainulainen P, Revonsuo A. *Neuropsychologia*. 2009 Nov; 47(13):2891–99.
21. "Neural correlates of temporality: Default mode variability and temporal awareness." Lloyd D. *Consciousness and Cognition*. 2011 Mar 17.
22. "Working memory capacity for spoken sentences decreases with adult ageing: Recall of fewer but not smaller chunks in older adults." Gilchrist A. L., Cowan N, Naveh-Benjamin M. *Memory*. 2008 Oct; 16(7):773–87.
23. "Neuroimaging analyses of human working memory." Smith E. E., Jonides J. *Proceedings of the National Academy of Sciences of the United States of America*. 1998 Sep 29; 95(20):12061–68.
24. "Working memory capacity for spoken sentences decreases with adult ageing: Recall of fewer but not smaller chunks in older adults." Gilchrist A. L., Cowan N, Naveh-Benjamin M. *Memory*. 2008 Oct; 16(7):773–87.
25. "Linguistically mediated visual search: The critical role of speech rate." Gibson B. S., Eberhard K. M., Bryant T. A. *Psychonomic Bulletin and Review*. 2005 Apr; 12(2):276–81.
26. "The influence of voice volume, pitch, and speech rate on progressive relaxation training: Application of methods from speech pathology and audiology." Knowlton G. E., Larkin K. T. *Applied Psychophysiology and Biofeedback*. 2006 Jun; 31(2):173–85.
27. "Synchronized brain activity during rehearsal and short-term memory

disruption by irrelevant speech is affected by recall mode." Kopp F, Schröger E, Lipka S. *International Journal of Psychophysiology*. 2006 Aug; 61(2):188–203.

28. "Irrelevant speech effects and statistical learning." Neath I, Guérard K, Jalbert A, Bireta T. J., Surprenant A. M. *Quarterly Journal of Experimental Psychology* (Colchester). 2009 Aug; 62(8):1551–59.
29. "Effects of target-masker contextual similarity on the multimasker penalty in a three-talker diotic listening task." Iyer N, Brungart D. S., Simpson B. D. *Journal of the Acoustical Society of America*. 2010 Nov; 128(5):2998–3010.
30. "Cross-modal distraction by background speech: What role for meaning?" Marsh J. E., Jones D. M. *Noise Health*. 2010 Oct–Dec; 12(49):210–16.
31. "Effects of road traffic noise and irrelevant speech on children's reading and mathematical performance." Ljung R, Sörqvist P, Hygge S. *Noise Health*. 2009 Oct–Dec; 11(45):194–98.
32. "Mind does really matter: Evidence from neuroimaging studies of emotional self-regulation, psychotherapy, and placebo effect." Beauregard M. *Progress in Neurobiology*. 2007 Mar; 81(4):218–36. Epub 2007 Feb 9.
33. "Embodied cognition and beyond: Acting and sensing the body." Borghi A. M., Cimatti F. *Neuropsychologia*. 2010 Feb; 48(3):763–73. "Voice: A pathway to consciousness as social contact to oneself." Bertau M. C. *Integrative Psychological and Behavioral Science*. 2008 Mar; 42(1):92–113.
34. "The voice of self-control: Blocking the inner voice increases impulsive responding." Tullett A. M., Inzlicht M. *Acta Psychologica* (Amsterdam). 2010 Oct; 135(2):252–56.
35. "The phenomena of inner experience." Heavey C. L., Hurlburt R. T. *Consciousness and Cognition*. 2008 Sep; 17(3):798–810.
36. "Private speech in adolescents." Kronk C. M. *Adolescence*. 1994 Winter; 29(116):781–804.
37. "Right hemispheric self-awareness: A critical assessment." Morin A. *Consciousness and Cognition*. 2002 Sep; 11(3):396–401.
38. "Self-awareness and the left inferior frontal gyrus: Inner speech use during self-related processing." Morin A, Michaud J. *Brain Research Bulletin*. 2007 Nov 1; 74(6):387–96. Epub 2007 Jul 5.
39. "Having a word with yourself: Neural correlates of self-criticism and self-

reassurance." Longe O, Maratos F. A., Gilbert P, Evans G, Volker F, Rockliff H, Rippon G. *NeuroImage*. 2010 Jan 15; 49(2):1849–56.

40. "Living with the anorexic voice: A thematic analysis." Tierney S, Fox J. R. *Psychology and Psychotherapy*. 2010 Sep; 83(Pt 3):243–54.
41. "Perceptions of self-concept and self-presentation by procrastinators: Further evidence." Ferrari J. R., Díaz-Morales J. F. *Spanish Journal of Psychology*. 2007 May; 10(1):91–96.
42. "Dynamic interactions between neural systems underlying different components of verbal working memory." Gruber O, Müller T, Falkai P. *Journal of Neural Transmission*. 2007; 114(8):1047–50.
43. "The effects of mindfulness-based stress reduction therapy on mental health of adults with a chronic medical disease: A meta-analysis." Bohlmeijer E, Prenger R, Taal E, Cuijpers P. *Journal of Psychosomatic Research*. 2010 Jun; 68(6):539–44.
44. "Inner speech as a retrieval aid for task goals: The effects of cue type and articulatory suppression in the random task cuing paradigm." Miyake A, Emerson M. J., Padilla F, Ahn J. C. *Acta Psychologica* (Amsterdam). 2004 Feb–Mar; 115(2–3):123–42.
45. "The phenomena of inner experience." Heavey C. L., Hurlburt R. T. *Consciousness and Cognition*. 2008 Sep; 17(3):798–810.
46. "Studying the effects of self-talk on thought content with male adult tennis players." Latinjak A. T., Torregrosa M, Renom J. *Perceptual and Motor Skills*. 2010 Aug; 111(1):249–60. "Effects of instructional and motivational self-talk on the vertical jump." Tod D. A., Thatcher R, McGuigan M, Thatcher J. *Journal of Strength and Conditioning Research*. 2009 Jan; 23(1):196–202.
47. "The dangers of feeling like a fake." De Vries M. F. *Harvard Business Review*. 2005 Sep; 83(9):108–16, 159.
48. "Interpretation of self-talk and post-lecture affective states of higher education students: A self-determination theory perspective." Oliver E. J., Markland D, Hardy J. *British Journal of Educational Psychology*. 2010 Jun; 80(Pt 2):307–23.
49. "Using self-talk to enhance career satisfaction and performance." White S. J. *American Journal of Health-System Pharmacy*. 2008 Mar 15; 65(6):514, 516, 519.

50. "Performance effects and subjective disturbance of speech in acoustically different office types—A laboratory experiment." Haka M, Haapakangas A, Keränen J, Hakala J, Keskinen E, Hongisto V. *Indoor Air*. 2009 Dec; 19(6):454–67.
51. "Neural activity when people solve verbal problems with insight." Jung-Beeman M, Bowden E. M., Haberman J, Frymiare J. L., Arambel-Liu S, Greenblatt R, Reber P. J., Kounios J. *PLoS Biology*. 2004 Apr; 2(4):E97.
52. "'Aha!': The neural correlates of verbal insight solutions." Aziz-Zadeh L, Kaplan J. T., Iacoboni M. *Human Brain Mapping*. 2009 Mar; 30(3):908–16. "Neural correlates of the 'Aha! reaction.'" Luo J, Niki K, Phillips S. *Neuroreport*. 2004 Sep 15; 15(13):2013–17.
53. "From alpha to gamma: Electrophysiological correlates of meditation-related states of consciousness." Fell J, Axmacher N, Haupt S. *Medical Hypotheses*. 2010 Aug; 75(2):218–24.
54. "Coherence in consciousness: Paralimbic gamma synchrony of self-reference links conscious experiences." Lou H. C., Gross J, Biermann-Ruben K, Kjaer T. W., Schnitzler A. *Human Brain Mapping*. 2010 Feb; 31(2):185–92.
55. "Listening to the sound of silence: Disfluent silent pauses in speech have consequences for listeners." MacGregor L. J., Corley M, Donaldson D. I. *Neuropsychologia*. 2010 Dec; 48(14):3982–92. Epub 2010 Oct 13.
56. "Neural activity in speech-sensitive auditory cortex during silence." Hunter M. D., Eickhoff S. B., Miller T. W., Farrow T. F., Wilkinson I. D., Woodruff P. W. *Proceedings of the National Academy of Sciences of the United States of America*. 2006 Jan 3; 103(1):189–94.
57. "The predicting brain: Unconscious repetition, conscious reflection and therapeutic change." Pally R. *International Journal of Psychoanalysis*. 2007 Aug; 88(Pt 4):861–81.
58. "Brain, conscious experience and the observing self." Baars B. J., Ramsøy T. Z., Laureys S. *Trends in Neurosciences*. 2003 Dec; 26(12):671–75.
59. "Self-rumination, self-reflection, and depression: Self-rumination counteracts the adaptive effect of self-reflection." Takano K, Tanno Y. *Behaviour Research and Therapy*. 2009 Mar; 47(3):260–64.
60. "Insight, rumination, and self-reflection as predictors of well-being." Harrington R, Loffredo D. A. *Journal of Psychology*. 2011 Jan–Feb; 145(1):39–57.

61. "Measuring mindfulness: Pilot studies with the Swedish versions of the Mindful Attention Awareness Scale and the Kentucky Inventory of Mindfulness Skills." Hansen E, Lundh L. G., Homman A, Wångby-Lundh M. *Cognitive Behaviour Therapy*. 2009; 38(1):2–15.
62. "Watching my mind unfold versus yours: An fMRI study using a novel camera technology to examine neural differences in self-projection of self versus other perspectives." St Jacques P. L., Conway M. A., Lowder M. W., Cabeza R. *Journal of Cognitive Neuroscience*. 2011 Jun; 23(6):1275–84.

Chapter 5: The Language of Cooperation (pages 77–86)

1. "Rules of social exchange: Game theory, individual differences and psychopathology." Wischniewski J, Windmann S, Juckel G, Brüne M. *Neuroscience and Biobehavioral Reviews*. 2009 Mar; 33(3):305–13.
2. "Long-term social bonds promote cooperation in the iterated prisoner's dilemma." St-Pierre A, Larose K, Dubois F. *Proceedings of the Royal Society of London: Series B, Biological Sciences*. 2009 Dec 7; 276(1676):4223–28.
3. "Cooperation within and among species." Sachs J. L. *Journal of Evolutionary Biology*. 2006 Sep; 19(5):1415-8; discussion 1426–36.
4. "From quorum to cooperation: Lessons from bacterial sociality for evolutionary theory." Lyon P. *Studies in History and Philosophy of Biological and Biomedical Sciences*. 2007 Dec; 38(4):820–33.
5. "Behavioural and community ecology of plants that cry for help." Dicke M. *Plant, Cell, and Environment*. 2009 Jun; 32(6):654–65.
6. "The evolutionary context for herbivore-induced plant volatiles: Beyond the 'cry for help.'" Dicke M, Baldwin I. T. *Trends in Plant Science*. 2010 Mar; 15(3):167–75. Epub 2010 Jan 4.
7. "Recent advances and emerging trends in plant hormone signalling." Santner A, Estelle M. *Nature*. 2009 Jun 25; 459(7250):1071–78.
8. "New evidence for a multi-functional role of herbivore-induced plant volatiles in defense against herbivores." Rodriguez-Saona C. R., Frost C. J. *Plant Signaling and Behavior*. 2010 Jan; 5(1):58–60.
9. "Playing charades in the fMRI: Are mirror and/or mentalizing areas involved in gestural communication?" Schippers M. B., Gazzola V, Goebel R, Keysers C. *PLoS One*. 2009 Aug 27; 4(8):e6801.
10. "Cooperation of different neuronal systems during hand sign recogni-

tion." Nakamura A, Maess B, Knösche T. R., Gunter T. C., Bach P, Friederici A. D. *NeuroImage*. 2004 Sep; 23(1):25–34.

11. "When your errors make me lose or win: Event-related potentials to observed errors of cooperators and competitors." Koban L, Pourtois G, Vocat R, Vuilleumier P. *Society for Neuroscience*. 2010; 5(4):360–74.
12. "Mirror neuron system involvement in empathy: A critical look at the evidence." Baird A. D., Scheffer I. E., Wilson S. J. *Society for Neuroscience*. 2011 Jan 10:1–9.
13. "Sociophysiology: Basic processes of empathy." Haker H, Schimansky J, Rössler W. *Neuropsychiatry*. 2010; 24(3):151–60.
14. "Neural activity predicts attitude change in cognitive dissonance." Van Veen V, Krug M. K., Schooler J. W., Carter C. S. *Nature Neuroscience*. 2009 Nov; 12(11):1469–74.
15. "The neural basis of rationalization: Cognitive dissonance reduction during decision-making." Jarcho J. M., Berkman E. T., Lieberman M. D. *Social Cognitive and Affective Neuroscience*. 2011 Sep; 6(4):460–67.
16. "Speaker-listener neural coupling underlies successful communication." Stephens G. J., Silbert L. J., Hasson U. *Proceedings of the National Academy of Sciences of the United States of America*. 2010 Aug 10; 107(32):14425–30.
17. "Effects of language intensity similarity on perceptions of credibility, relational attributions, and persuasion." Aune R. K., Kikuchi T. *Journal of Language and Social Psychology.* 1993 12:224.
18. "Language style matching predicts relationship initiation and stability." Ireland M. E., Slatcher R. B., Eastwick P. W., Scissors L. E., Finkel E. J., Pennebaker J. W. *Psychological Science*. 2011 Jan 1; 22(1):39–44.
19. "Human mimicry." Chartrand T. L., Van Baaren R. *Advances in Experimental Social Psychology.* 2009 41:219–74.
20. "Where is the love? The social aspects of mimicry." Van Baaren R, Janssen L, Chartrand T. L., Dijksterhuis A. *Philosophical Transactions of the Royal Society of London: Section B, Biological Sciences*. 2009 Aug 27; 364(1528):2381–89.
21. "Using nonconscious behavioral mimicry to create affiliation and rapport." Lakin J. L., Chartrand T. L. *Psychological Science*. 2003 Jul; 14(4):334–39.
22. "Mimicry for money: Behavioral consequences of imitation." Van Baaren

R. B., Holland R. W., Steenaert B, Van Knippenberg A. *Journal of Experimental Social Psychology*. 2003 39:393–98.

23. "Linguistic style matching and negotiation outcome." Taylor P. J., Thomas S. *Negotiation and Conflict Management Research*. 2008 1:263–81.
24. "The power of simulation: Imagining one's own and other's behavior." Decety J, Grèzes J. *Brain Research*. 2006 Mar 24; 1079(1):4–14.
25. "Impact of interactivity on identification with characters in fiction." Soto-Sanfiel M. T., Aymerich-Franch L, Ribes Guàrdia F. X. *Psicothema*. 2010 Nov; 22(4):822–27.
26. "The neural substrates of cognitive empathy." Preston S. D., Bechara A, Damasio H, Grabowski T. J., Stansfield R. B., Mehta S, Damasio A. R. *Society for Neuroscience*. 2007; 2(3–4):254–75.
27. "Social neuroscience: Mirror neurons recorded in humans." Keysers C, Gazzola V. *Current Biology* 27 2010; Apr 8 (8):353–54.
28. "The effect of empathy on accuracy of behavior prediction in social exchange situation." Tanida S, Yamagishi T. *Shinrigaku Kenkyu*. 2004 Feb; 74(6):512–20.
29. "Psychophysiology of neural, cognitive and affective integration: fMRI and autonomic indicants." Critchley H. D. *International Journal of Psychophysiology*. 2009 Aug; 73(2):88–94.
30. "Evolved altruism, strong reciprocity, and perception of risk." Tucker W. T., Ferson S. *Annals of the New York Academy of Sciences*. 2008 Apr; 1128:111–20.
31. "Evolution of cooperation and altruistic punishment when retaliation is possible." Janssen M. A., Bushman C. *Journal of Theoretical Biology*. 2008 Oct 7; 254(3):541–45.
32. "Am 'I' more important than 'we'? Couples' word use in instant messages." Slatcher R. B., Vazire S, Pennebaker J. W. *Personal Relationships* 2008 15:407–24.
33. "Loving-kindness meditation increases social connectedness." Hutcherson C. A., Seppala E. M., Gross J. J. *Emotion*. 2008 Oct; 8(5):720–4. "Social neuroscience, empathy, brain integration, and neurodevelopmental disorders." Harris J. C. Physiology and Behavior. 2003 Aug; 79(3):525–31.
34. "Winners don't punish." Dreber A, Rand D. G., Fudenberg D, Nowak M. A. *Nature*. 2008 Mar 20; 452(7185):348–51.

35. "Skills development for conflict transformation: A training manual on understanding conflict, negotiation and mediation." United Nations Conflict Management Project. http://unpan1.un.org/intradoc/groups/public/documents/un/unpan001363.pdf.
36. "Partner choice creates competitive altruism in humans." Barclay P, Willer R. *Proceedings of the Royal Society of London: Series B, Biological Sciences.* 2007 Mar 7; 274(1610):749–53.

Chapter 6: The Language of Trust (pages 87–100)

1. "Common neural mechanisms for the evaluation of facial trustworthiness and emotional expressions as revealed by behavioral adaptation." Engell A. D., Todorov A, Haxby J. V. *Perception.* 2010; 39(7):931–41.
2. "Eyes are on us, but nobody cares: Are eye cues relevant for strong reciprocity?" Fehr E, Schneider F. *Proceedings of the Royal Society of London: Series B, Biological Sciences.* 2010 May 7; 277(1686):1315–23.
3. "Evaluating faces on trustworthiness: An extension of systems for recognition of emotions signaling approach/avoidance behaviors." Todorov A. *Annals of the New York Academy of Sciences.* 2008 Mar; 1124:208–24.
4. "Cues of being watched enhance cooperation in a real-world setting." Bateson M, Nettle D, Roberts G. *Biology Letters.* 2006 Sep 22; 2(3):412–14.
5. " 'Big Brother' eyes inspire police crime crackdown campaign." Press release, Newcastle University. September 6, 2006.
6. BBC News. December 8, 2005. http://news.bbc.co.uk/2/hi/uk_news/england/derbyshire/4511674.stm.
7. "Effects of anonymity on antisocial behavior committed by individuals." Nogami T, Takai J. *Psychological Reports.* 2008 Feb; 102(1):119–30.
8. "Neural bases of eye and gaze processing: The core of social cognition." Itier R. J., Batty M. *Neuroscience and Biobehavioral Reviews.* 2009 Jun; 33(6):843–63.
9. " 'Did you call me?' 5-month-old infant's own name guides their attention." Parise E, Friederici A. D., Striano T. *PLoS One.* 2010 Dec 3; 5(12):e14208.
10. "Facing the gaze of others." George N, Conty L. *Clinical Neurophysiology.* 2008 Jun; 38(3):197–207.
11. " Seeing direct and averted gaze activates the approach-avoidance moti-

vational brain systems "Hietanen J. K., Leppänen J. M., Peltola M. J., Linna-Aho K, Ruuhiala H. J. *Neuropsychologia*. 2008; 46(9):2423–30.

12. "Tell-tale eyes: Children's attribution of gaze aversion as a lying cue." Einav S, Hood B. M. *Developmental Psychology*. 2008 Nov; 44(6):1655–67.
13. "Fear and avoidance of eye contact in social anxiety disorder." Schneier F. R., Rodebaugh T. L., Blanco C, Lewin H, Liebowitz M. R. *Comprehensive Psychiatry*. 2011 Jan–Feb;52(1):81–87.
14. "Cultural display rules drive eye gaze during thinking." McCarthy A, Lee K, Itakura S, Muir D. W. *Journal of Cross-Cultural Psychology*. 2006 Nov; 37(6):717–22.
15. Ekman P. *Emotions Revealed*. Holt, 2007.
16. If you want to train yourself to read facial expressions more accurately, read his book *Emotions Revealed* and visit his website (http://www.paulekman.com) where you can take an online training course for recognizing micro-expressions that occur in less than a second. The training is used by therapists, law enforcement officers, business managers, and salespeople because it helps them to communicate more effectively with others. The television series *Lie to Me* is based on Ekman's work.
17. "Is eye to eye contact really threatening and avoided in social anxiety?—An eye-tracking and psychophysiology study." Wieser M. J., Pauli P, Alpers G. W., Mühlberger A. *Journal of Anxiety Disorders*. 2009 Jan; 23(1):93–103.
18. "Amygdala activation predicts gaze toward fearful eyes." Gamer M, Büchel C. *Journal of Neuroscience*. 2009 Jul 15; 29(28):9123–26.
19. "Brain response to a humanoid robot in areas implicated in the perception of human emotional gestures." Chaminade T, Zecca M, Blakemore S. J., Takanishi A, Frith C. D., Micera S, Dario P, Rizzolatti G, Gallese V, Umiltà M. A. *PLoS One*. 2010 Jul 21; 5(7):e11577.
20. "The Duchenne smile: Emotional expression and brain physiology II." Ekman P, Davidson R. J., Friesen W. V. *Journal of Personality and Social Psychology*. 1990 Feb; 58(2):342–53.
21. "What makes Mona Lisa smile?" Kontsevich L. L., Tyler C. W. *Vision Research*. 2004; 44(13):1493–98.
22. "Facial and emotional reactions to Duchenne and non-Duchenne smiles." Surakka V, Hietanen J. K. *International Journal of Psychophysiology*. 1998 Jun; 29(1):23–33.

23. "Duchenne smile, emotional experience, and autonomic reactivity: A test of the facial feedback hypothesis." Soussignan R. *Emotion*. 2002 Mar; 2(1):52–74.
24. "Why are smiles contagious? An fMRI study of the interaction between perception of facial affect and facial movements." Wild B, Erb M, Eyb M, Bartels M, Grodd W. *Psychiatry Research*. 2003 May 1; 123(1):17–36.
25. "What's in a 'smile'? Intra-operative observations of contralateral smiles induced by deep brain stimulation." Okun M. S., Bowers D, Springer U, Shapira N. A., Malone D, Rezai A. R., Nuttin B, Heilman K. M., Morecraft R. J., Rasmussen S. A., Greenberg B. D., Foote K. D., Goodman W. K. *Neurocase*. 2004 Aug; 10(4):271–79.
26. "What's in a smile? Maternal brain responses to infant facial cues." Strathearn L, Li J, Fonagy P, Montague P. R. *Pediatrics*. 2008 Jul; 122(1):40–51.
27. "Anticipatory smiling: linking early affective communication and social outcome." Parlade M. V., Messinger D. S., Delgado C. E., Kaiser M. Y., van Hecke A. V., Mundy P. C. *Infant Behavior and Development*. 2009 Jan; 32(1):33–43.
28. "Smile production in older infants: The importance of a social recipient for the facial signal." Jones S. S., Raag T. *Child Development*. 1989 Aug; 60(4):811–18.
29. "Love hurts: An fMRI study." Cheng Y, Chen C, Lin C. P., Chou K. H, Decety J. *NeuroImage*. 2010 Jun; 51(2):923–29.
30. "What's in a smile? Maternal brain responses to infant facial cues." Strathearn L, Li J, Fonagy P, Montague P. R. *Pediatrics*. 2008 Jul; 122(1):40–51. "Regulation of the neural circuitry of emotion by compassion meditation: Effects of meditative expertise." Lutz A, Brefczynski-Lewis J, Johnstone T, Davidson R.J. *PLoS One*. 2008 Mar 26; 3(3):e1897.

Chapter 7: Inner Values (pages 103–19)

1. "Affirmation of personal values buffers neuroendocrine and psychological stress responses." Creswell J. D., Welch W. T., Taylor S. E., Sherman D. K., Gruenewald T. L., Mann T. *Psychological Science*. 2005 Nov; 16(11):846–51.
2. "Do messages about health risks threaten the self? Increasing the acceptance of threatening health messages via self-affirmation." Sherman D. K.,

Nelson L. D., Steele C. M. *Personality and Social Psychology Bulletin.* 2000 26:1046–58. "The cessation of rumination through self-affirmation." Koole S. L., Smeets K, Van Knippenberg A, Dijksterhuis A. *Journal of Personality and Social Psychology.* 1999 77:111–25.

3. "Personal values and pain tolerance: Does a values intervention add to acceptance?" Branstetter-Rost A, Cushing C, Douleh T. *Journal of Pain.* 2009 Aug; 10(8):887–92.
4. "Getting value from value." Kanter R. M. *Harvard Business Review* (blog). 2010 Jun 14.
5. Ibid.
6. "Food choice motives and bread liking of consumers embracing hedonistic and traditional values." Pohjanheimo T, Paasovaara R, Luomala H, Sandell M. *Appetite.* 2010 Feb; 54(1):170–80.
7. "Absolute versus relative values: Effects on medical decisions and personality of patients and physicians." Neumann J. K., Olive K. E., McVeigh S. D. *Southern Medical Journal.* 1999 Sep; 92(9):871–76.
8. "Genetic and environmental influences on girls' and boys' gender-typed and gender-neutral values." Knafo A, Spinath F. M. Developmental Psychology. 2011 May; 47(3):726–31. "Phenotypic, genetic, and environmental properties of the portrait values questionnaire." Schermer J. A., Feather N. T., Zhu G, Martin N. G. *Twin Research and Human Genetics.* 2008 Oct; 11(5):531–37.
9. "The significance of task significance: Job performance effects, relational mechanisms, and boundary conditions." Grant A. M. *Journal of Applied Psychology.* 2008 Jan; 93(1):108–24. "Personal values as a mediator between parent and peer expectations and adolescent behaviors." Padilla-Walker L. M., Carlo G. *Journal of Family Psychology.* 2007 Sep; 21(3):538–41. "Social orientation: Problem behavior and motivations toward interpersonal problem solving among high risk adolescents." Kuperminc G. P., Allen J. P. *Journal of Youth and Adolescence.* 2001 Oct; 30(5):597–622.
10. "Neural basis of individualistic and collectivistic views of self." Chiao J. Y., Harada T, Komeda H, Li Z, Mano Y, Saito D, Parrish T. B., Sadato N, Iidaka T. *Human Brain Mapping.* 2009 Sep; 30(9):2813–20.
11. "Cultural influences on neural substrates of attentional control." Hedden

T, Ketay S, Aron A, Markus H. R., Gabrieli J. D. *Psychological Science*. 2008 Jan; 19(1):12–17.

12. "High income improves evaluation of life but not emotional well-being." Kahneman D, Deaton A. *Proceedings of the National Academy of Sciences of the United States of America*. 2010 Sep 21; 107(38):16489–93.
13. "Wealth and happiness across the world: Material prosperity predicts life evaluation, whereas psychosocial prosperity predicts positive feeling." Diener E, Ng W, Harter J, Arora R. *Journal of Personality and Social Psychology*. 2010 Jul; 99(1):52–61.
14. "Money giveth, money taketh away: The dual effect of wealth on happiness." Quoidbach J, Dunn E. W., Petrides K. V., Mikolajczak M. *Psychological Science*. 2010 Jun; 21(6):759–63.
15. "Near death experiences, cognitive function and psychological outcomes of surviving cardiac arrest." Parnia S, Spearpoint K, Fenwick P. B. *Resuscitation*. 2007 Aug; 74(2):215–21.
16. "Greed, death, and values: From terror management to transcendence management theory." Cozzolino P. J., Staples A. D., Meyers L. S., Samboceti J. *Personality and Social Psychology Bulletin*. 2004 Mar; 30(3):278–92.
17. "Nurses' professional and personal values." Rassin M. *Nursing Ethics*. 2008 Sep; 15(5):614–30.
18. "Burnout and nurses' personal and professional values." Altun I. *Nursing Ethics*. 2002 May; 9(3):269–78.
19. "Demands, values, and burnout: Relevance for physicians." Leiter M. P., Frank E, Matheson T. J. *Canadian Family Physician*. 2009 Dec; 55(12):1224–25, 1225.e1–6.
20. "Nursing values and a changing nurse workforce: Values, age, and job stages." McNeese-Smith D. K., Crook M. *Journal of Nursing Administration*. 2003 May; 33(5):260–70.
21. "The power of values." Levin R. P. *Journal of the American Dental Association*. 2003 Nov; 134(11):1520–21.
22. "Spirituality in higher education: A national study of college students' search for meaning and purpose." http://www.spirituality.ucla.edu.

Chapter 8: Twelve Steps to Intimacy, Cooperation, and Trust (pages 121–45)

1. "Stress overload: A new diagnosis." Lunney M. *International Journal of Nursing Terminologies and Classifications*. 2006 Oct–Dec; 17(4):165–75.
2. "Short-term meditation training improves attention and self-regulation." Tang Y. Y., Ma Y, Wang J, Fan Y, Feng S, Lu Q, Yu Q, Sui D, Rothbart M. K., Fan M, Posner M. I. *Proceedings of the National Academy of Sciences of the United States of America*. 2007 Oct 23; 104(43):17152–56.
3. "An investigation of brain processes supporting meditation." Baerentsen K. B., Stødkilde-Jørgensen H, Sommerlund B, Hartmann T, Damsgaard-Madsen J, Fosnaes M, Green A. C. *Cognitive Processing*. 2010 Feb; 11(1):57–84.
4. "Exploring co-meditation as a means of reducing anxiety and facilitating relaxation in a nursing school setting." Malinski V. M., Todaro-Franceschi V. *Journal of Holistic Nursing*. 2011 Feb 28.
5. Tolle E. *Gateways to Now*. Simon and Schuster Audio, 2003.
6. "Object-based attention: Shifting or uncertainty?" Drummond L, Shomstein S. *Attention, Perception, and Psychophysics*. 2010 Oct; 72(7):1743–55.
7. "'Thinking about not-thinking': Neural correlates of conceptual processing during Zen meditation." Pagnoni G, Cekic M, Guo Y. *PLoS One*. 2008 Sep 3; 3(9):e3083.
8. "Age effects on gray matter volume and attentional performance in Zen meditation." Pagnoni G, Cekic M. *Neurobiology of Aging*. 2007 Oct; 28(10):1623–27.
9. Fredrickson B. *Positivity*. Three Rivers Press, 2009.
10. "The role of positivity and connectivity in the performance of business teams: A nonlinear dynamics model." Losada M, Heaphy E. *American Behavioral Scientist*. 2004; 47 (6):740–65.
11. Gottman J. *What Predicts Divorce?: The Relationship Between Marital Processes and Marital Outcomes*. Psychology Press, 1993.
12. "Optimal and normal affect balance in psychotherapy of major depression: Evaluation of the balanced states of mind model." Schwartz R. M., Reynolds C. F., Thase M. E., Frank E, Fasiczka A. L., Haaga D. A. F. *Behavioral and Cognitive Psychotherapy*. 2002 Oct; 30(4):439–50.

13. "Patient-provider communication and low-income adults: Age, race, literacy, and optimism predict communication satisfaction." Jensen J. D., King A. J., Guntzviller L. M., Davis L. A. *Patient and Educational Counseling*. 2010 Apr; 79(1):30–35.
14. "Seeing future success: Does imagery perspective influence achievement motivation?" Vasquez N. A., Buehler R. *Personality and Social Psychology Bulletin*. 2007 Oct; 33(10):1392–405.
15. "Mental imagery and emotion in treatment across disorders: Using the example of depression." Holmes E. A., Lang T. J., Deeprose C. *Cognitive Behaviour Therapy*. 2009 Aug; 20:1.
16. "Positive interpretation training: Effects of mental imagery versus verbal training on positive mood." Holmes E. A., Mathews A, Dalgleish T, Mackintosh B. *Behavior Therapy*. 2006 Sep; 37(3):237–47.
17. "Mental imagery as an emotional amplifier: Application to bipolar disorder." Holmes E. A., Geddes J. R., Colom F, Goodwin G. M. *Behaviour Research and Therapy*. 2008 Dec; 46(12):1251–58.
18. "Giving off a rosy glow: The manipulation of an optimistic orientation." Fosnaugh J, Geers A. L., Wellman J. A. *Journal of Social Psychology*. 2009 Jun; 149(3):349–64.
19. "Treatment of childhood memories: Theory and practice." Arntz A, Weertman A. *Behaviour Research and Therapy*. 1999 Aug; 37(8):715–40.
20. Seligman M. *Learned Optimism*. Free Press, 1997.
21. "Enhancing well-being and alleviating depressive symptoms with positive psychology interventions: A practice-friendly meta-analysis." Sin N. L., Lyubomirsky S. *Journal of Clinical Psychology*. 2009 May; 65(5):467–87.
22. "Deviance among young Italians: Investigating the predictive strength of value systems." Froggio G, Lori M. *International Journal of Offender Therapy and Comparative Criminology*. 2010 Aug; 54(4):581–96.
23. "Facial expressions, their communicatory functions and neuro-cognitive substrates." Blair R. J. *Philosophical Transactions of the Royal Society of London: Series B, Biological Sciences*. 2003 Mar 29; 358(1431):561–72.
24. "The facial expression says more than words: Is emotional 'contagion' via facial expression the first step toward empathy?" Sonnby-Borgström M. *Lakartidningen*. 2002 Mar 27; 99(13):1438–42.
25. "Effectiveness of training in negative thought reduction and positive

thought increment in reducing thought-produced distress." Dua J, Price I. *Journal of Genetic Psychology.* 1993 Mar;154(1):97–109.

26. Nöth W. *Handbook of Semiotics.* Indiana University Press, 1990.
27. "The lived experience of contentment: A study using the Parse research method." Parse R. R. *Nursing Science Quarterly.* 2001 Oct; 14(4):330–38.
28. "The eye contact effect: Mechanisms and development." Senju A, Johnson M. H. *Trends in Cognitive Sciences.* 2009 Mar; 13(3):127–34.
29. "Oxytocin enhances amygdala-dependent, socially reinforced learning and emotional empathy in humans." Hurlemann R, Patin A, Onur O. A., Cohen M. X., Baumgartner T, Metzler S, Dziobek I, Gallinat J, Wagner M, Maier W, Kendrick K. M. *Journal of Neuroscience.* 2010 Apr 7; 30(14):4999–5007. "Intranasal oxytocin increases positive communication and reduces cortisol levels during couple conflict." Ditzen B, Schaer M, Gabriel B, Bodenmann G, Ehlert U, Heinrichs M. *Biological Psychiatry.* 2009 May 1; 65(9):728–31.
30. "Become versed in reading faces." Ekman P. *Entrepreneur,* March 26, 2009.
31. "Self-objectification and compliment type: Effects on negative mood." Fea C. J., Brannon L. A. *Body Image.* 2006 Jun; 3(2):183–88.
32. http://www.mayoclinic.com/health/how-to-be-happy/MY01357.
33. " 'I've heard wonderful things about you': How patients compliment surgeons." Hudak P. L., Gill V. T., Aguinaldo J. P., Clark S, Frankel R. *Sociology of Health and Illness.* 2010 Jul; 32(5):777–97.
34. "Facial and vocal expressions of emotion." Russell J. A., Bachorowski J. A., Fernandez-Dols J. M. *Annual Review of Psychology.* 2003; 54:329–49.
35. " 'Worth a thousand words': Absolute and relative decoding of nonlinguistic affect vocalizations." Hawk S. T., Van Kleef G. A., Fischer A. H., Van der Schalk J. *Emotion.* 2009 Jun; 9(3):293–305.
36. "Perceptual cues in nonverbal vocal expressions of emotion." Sauter D. A., Eisner F, Calder A. J., Scott S. K. *Quarterly Journal of Experimental Psychology* (Colchester). 2010 Nov; 63(11):2251–72.
37. "Mapping emotions into acoustic space: The role of voice production." Patel S, Scherer K. R., Björkner E, Sundberg J. *Biological Psychiatry.* 2011 Apr; 87(1):93–98.
38. "Voice analysis during bad news discussion in oncology: Reduced pitch, decreased speaking rate, and nonverbal communication of empathy."

McHenry M, Parker P. A., Baile W. F., Lenzi R. *Supportive Care in Cancer.* 2011 May 15.

39. "Components of placebo effect: Randomised controlled trial in patients with irritable bowel syndrome." Kaptchuk T. J., Kelley J. M., Conboy L. A., Davis R. B., Kerr C. E., Jacobson E. E., Kirsch I, Schyner R. N., Nam B. H., Nguyen L. T., Park M, Rivers A. L., McManus C, Kokkotou E, Drossman D. A., Goldman P, Lembo A. J. *British Medical Journal.* 2008 May 3; 336(7651):999–1003.
40. "Leadership = communication? The relations of leaders' communication styles with leadership styles, knowledge sharing and leadership outcomes." De Vries R. E., Bakker-Pieper A, Oostenveld W. *Journal of Business Psychology.* 2010 Sep; 25(3):367–80.
41. "'It's not what you say, but how you say it': A reciprocal temporo-frontal network for affective prosody." Leitman D. I., Wolf D. H., Ragland J. D., Laukka P, Loughead J, Valdez J. N., Javitt D. C., Turetsky B. I., Gur R. C. *Frontiers in Human Neuroscience.* 2010 Feb 26; 4:19.
42. "Use of affective prosody by young and older adults." Dupuis K, Pichora-Fuller M. K. *Psychology and Aging.* 2010 Mar; 25(1):16–29.
43. "Responses of single neurons in monkey amygdala to facial and vocal emotions." Kuraoka K, Nakamura K. *Journal of Neurophysiology.* 2007 Feb; 97(2):1379–87.
44. "Comprehension of speeded discourse by younger and older listeners." Gordon M.S., Daneman M, Schneider B. A. *Experimental Aging Research.* 2009 Jul–Sep; 35(3):277–96.
45. "Celerity and cajolery: Rapid speech may promote or inhibit persuasion through its impact on message elaboration." Smith S. M., Shaffer, D. R. *Personality and Social Psychology Bulletin.* 1991 Dec; 17(6):663–69.
46. "The effect of rate control on speech rate and intelligibility of dysarthric speech." Van Nuffelen G, De Bodt M, Wuyts F, Van de Heyning P. *Folia Phoniatrica et Logopaedica.* 2009; 61(2):69–75. "Influences of rate, length, and complexity on speech disfluency in a single-speech sample in preschool children who stutter." Sawyer J, Chon H, Ambrose N. G. *Journal of Fluency Disorders.* 2008 Sep; 33(3):220–40.
47. "The influence of speech rate stereotypes and rate similarity or listeners' evaluations of speakers." Street R. L., Brady R. M., Putman W. B. *Journal of Language and Social Psychology.* 1983 Mar; (2):37–56.

48. "Are fast talkers more persuasive?" Dean J. *Psyblog*: http://www.spring.org.uk/2010/11/are-fast-talkers-more-persuasive.php.
49. "Influence of mothers' slower speech on their children's speech rate." Guitar B, Marchinkoski L. *Journal of Speech, Language, and Hearing Research*. 2001 Aug; 44(4):853–61.
50. "Voices of fear and anxiety and sadness and depression: The effects of speech rate and loudness on fear and anxiety and sadness and depression." Siegman A. W., Boyle S. *Journal of Abnormal Psychology*. 1993 Aug; 102(3):430–37; "The angry voice: Its effects on the experience of anger and cardiovascular reactivity." Siegman A. W., Anderson R. A., Berger T. *Psychosomatic Medicine*. 1990 Nov–Dec; 52(6):631–43.
51. "Feeling listened to: A lived experience of human becoming." Kagan P. N. *Nursing Science Quarterly*. January 2008 21: 59–67. Feeling understood: a melody of human becoming. Jonas-Simpson C. M. *Nursing Science Quarterly*. 2001 Jul; 14(3):222–30.
52. "What is the relationship between phonological short-term memory and speech processing?" Jacquemot C, Scott S. K. *Trends in Cognitive Sciences*. 2006 Nov; 10(11):480–86.
53. "Soliciting the patient's agenda: have we improved?" Marvel M. K., Epstein R. M., Flowers K, Beckman H. B. *Journal of the American Medical Association*. 1999 Jan 20; 281(3):283–87.

Chapter 9: Compassionate Communication (pages 147–62)

1. "Journaling about stressful events: Effects of cognitive processing and emotional expression." Ullrich P. M., Lutgendorf S. K. *Annals of Behavioral Medicine*. 2002 Summer; 24(3):244–50. "The effects of journaling for women with newly diagnosed breast cancer." Smith S, Anderson-Hanley C, Langrock A, Compas B. *Psycho-Oncology*. 2005 Dec; 14(12):1075–82.
2. "Moderators of cardiovascular reactivity to speech: Discourse production and group variations in blood pressure and pulse rate." Tardy C. H., Allen M. T. *International Journal of Psychophysiology*. 1998 Aug; 29(3):247–54.

Chapter 10: Compassionate Communication with Loved Ones (pages 165–81)

1. "Trust, variability in relationship evaluations, and relationship processes." Campbell L, Simpson J. A., Boldry J. G., Rubin H. *Journal of Personality and Social Psychology*. 2010 Jul; 99(1):14–31.
2. "Perceptions of conflict and support in romantic relationships: the role of attachment anxiety." Campbell L, Simpson J. A., Boldry J, Kashy D. A. *Journal of Personality and Social Psychology*. 2005 Mar; 88(3):510–31.
3. "Calibrating the sociometer: The relational contingencies of self-esteem." Murray S. L., Griffin D. W., Rose P, Bellavia G. M. *Journal of Personality and Social Psychology*. 2003 Jul; 85(1):63–84.
4. "Risk assessment as an evolved threat detection and analysis process". Blanchard D. C, Griebel G, Pobbe R, Blanchard R. J. *Neuroscience and Biobehavioral Reviews*. 2011 Mar; 35(4):991–98. "Mirror neurons, procedural learning, and the positive new experience: A developmental systems self psychology approach." Wolf N. S., Gales M, Shane E, Shane M. *Journal of the American Academy of Psychoanalysis*. 2000 Fall; 28(3):409–30.
5. "Neural activity to a partner's facial expression predicts self-regulation after conflict." Hooker C. I., Gyurak A, Verosky S. C., Miyakawa A, Ayduk O. *Biological Psychiatry*. 2010 Mar 1; 67(5):406–13.
6. "Older spouses' cortisol responses to marital conflict: Associations with demand/withdraw communication patterns." Heffner K. L., Loving T. J., Kiecolt-Glaser J. K., Himawan L. K., Glaser R, Malarkey W. B. *Journal of Behavioral Medicine*. 2006 Aug; 29(4):317–25.
7. "Conflict and collaboration in middle-aged and older couples: II. Cardiovascular reactivity during marital interaction." Smith T. W., Uchino B. N., Berg C. A., Florsheim P, Pearce G, Hawkins M, Henry N. J., Beveridge R. M., Skinner M. A., Ko K. J., Olsen-Cerny C. *Psychology and Aging*. 2009 Jun; 24(2):274–86.
8. "Hostile marital interactions, proinflammatory cytokine production, and wound healing." Kiecolt-Glaser J. K., Loving T. J., Stowell J. R., Malarkey W. B., Lemeshow S, Dickinson S. L., Glaser R. *Archives of General Psychiatry*. 2005 Dec; 62(12):1377–84.
9. "Marital behavior, oxytocin, vasopressin, and wound healing." Gouin J. P., Carter C. S., Pournajafi-Nazarloo H, Glaser R, Malarkey W. B., Loving

T. J., Stowell J, Kiecolt-Glaser J. K. *Psychoneuroendocrinology*. 2010 Aug; 35(7):1082–90.

10. "The role of mindfulness in romantic relationship satisfaction and responses to relationship stress." Barnes S, Brown K. W., Krusemark E, Campbell W. K., Rogge R. D. *Journal of Marital and Family Therapy*. 2007 Oct; 33(4):482–500.
11. "Resolving unfinished business: relating process to outcome." Greenberg L.S., Malcolm W. *Journal of Consulting and Clinical Psychology*. 2002 Apr; 70(2):406–16.
12. "Resolving 'unfinished business': Efficacy of experiential therapy using empty-chair dialogue." Paivio S. C., Greenberg L. S. *Journal of Consulting and Clinical Psychology*. 1995 Jun; 63(3):419–25.
13. Hanh T. *Present Moment, Wonderful Moment*. Parallax, 1990.
14. "How do I love thee? Let me count the words: The social effects of expressive writing." Slatcher R. B., Pennebaker J. W. *Psychological Science*. 2006 Aug; 17(8):660–64.
15. "Cognitive word use during marital conflict and increases in proinflammatory cytokines." Graham J. E., Glaser R, Loving T. J., Malarkey W. B., Stowell J. R., Kiecolt-Glaser J. K. *Health Psychology*. 2009 Sep; 28(5):621–30.
16. "The autonomic phenotype of rumination." Ottaviani C, Shapiro D, Davydov D. M., Goldstein I. B., Mills P. J. *International Journal of Psychophysiology*. 2009 Jun; 72(3):267–75.
17. "Impulsivity and schemas for a hostile world: Postdictors of violent behaviour." James M, Seager J. A. *International Journal of Offender Therapy and Comparative Criminology*. 2006 Feb; 50(1):47–56.
18. "Explication of interspousal criticality bias." Peterson K. M., Smith D. A., Windle C. R. *Behaviour Research and Therapy*. 2009 Jun; 47(6):478–86.
19. "Overperception of spousal criticism in dysphoria and marital discord." Smith D. A., Peterson K. M. *Behavior Therapy*. 2008 Sep; 39(3):300–312.
20. "Marital satisfaction, depression, and attributions: A longitudinal analysis." Fincham F. D., Bradbury T. N. *Journal of Personality and Social Psychology*. 1993 Mar; 64(3):442–52.
21. "Expressed emotion, perceived criticism and 10-year outcome of depression." Kronmüller K. T., Backenstrass M, Victor D, Postelnicu I, Schen-

kenbach C, Joest K, Fiedler P, Mundt C. *Psychiatry Research*. 2008 May 30; 159(1–2):50–5.

22. "To what does perceived criticism refer? Constructive, destructive, and general criticism." Peterson K. M., Smith D. A. *Journal of Family Psychology*. 2010 Feb; 24(1):97–100.
23. "Attachment style, excessive reassurance seeking, relationship processes, and depression." Shaver P. R., Schachner D. A., Mikulincer M. *Personality and Social Psychology Bulletin*. 2005 Mar; 31(3):343–59.
24. "Evolving knowledge of sex differences in brain structure, function, and chemistry." Cosgrove K. P., Mazure C. M., Staley J. K. *Biological Psychiatry*. 2007 Oct 15; 62(8):847–55. "Gender differences in cognitive functions." Weiss E. M., Deisenhammer E. A., Hinterhuber H, Marksteiner J. *Fortschr Neurol Psychiatr*. 2005 Oct; 73(10):587–95.
25. "No gender differences in brain activation during the N-back task: An fMRI study in healthy individuals." Schmidt H, Jogia J, Fast K, Christodoulou T, Haldane M, Kumari V, Frangou S. *Human Brain Mapping*. 2009 Nov; 30(11):3609–15. "On sex/gender related similarities and differences in fMRI language research." Kaiser A, Haller S, Schmitz S, Nitsch C. *Brain Research Reviews*. 2009 Oct; 61(2):49–59. Epub 2009 May 4.
26. "Top 10 myths about the brain." Helmuth L. Smithsonian.com, May 20, 2011.

Chapter 11: Compassionate Communication in the Workplace (pages 183–96)

1. "Face value: Amygdala response reflects the validity of first impressions." Rule N. O., Moran J. M., Freeman J. B., Whitfield-Gabrieli S, Gabrieli J. D., Ambady N. *NeuroImage*. 2011 Jan 1; 54(1):734–41.
2. "Leader-follower values congruence: Are socialized charismatic leaders better able to achieve it?" Brown M. E., Treviño L. K. *Journal of Applied Psychology*. 2009 Mar; 94(2):478–90.
3. "Managing oneself." Drucker, P. R. *The Best of Harvard Business Review*. 1999; reprinted January 2005.
4. "Are you wasting your time on values statements?" Goldsmith, M. *Huffpost Business*. July 4, 2009. http://www.huffingtonpost.com/marshall-goldsmith/values-you-see-in-action_b_231131.html.

5. Goldsmith M. In *Leadership Coaching*, Passmore J, ed. Kogan Page, 2010.
6. "The bright-side and the dark-side of CEO personality: Examining core self-evaluations, narcissism, transformational leadership, and strategic influence." Resick C. J., Whitman D. S., Weingarden S. M., Hiller N. J. *Journal of Applied Psychology*. 2009 Nov; 94(6):1365–81
7. "A study of leadership behaviors among chairpersons in allied health programs." Firestone D. T. *Journal of Allied Health*. 2010 Spring; 39(1):34–42.
8. "Principals' transformational leadership and teachers' collective efficacy." Dussault M, Payette D, Leroux M. *Psychological Reports*. 2008 Apr; 102(2):401–10.
9. "An examination of 'nonleadership': From laissez-faire leadership to leader reward omission and punishment omission." Hinkin T. R., Schriesheim C. A. *Journal of Applied Psychology*. 2008 Nov; 93(6):1234–48.
10. "The destructiveness of laissez-faire leadership behavior." Skogstad A, Einarsen S, Torsheim T, Aasland M. S., Hetland H. *Journal of Occupational Health Psychology*. 2007 Jan; 12(1):80–92.
11. "Breakthrough bargaining." Kolb D. M., Williams J. *Harvard Business Review*. 2001 Feb; 79(2):88–97, 156.
12. "Leadership = communication? The relations of leaders' communication styles with leadership styles, knowledge sharing and leadership outcomes." De Vries R. E., Bakker-Pieper A, Oostenveld W. *Journal of Business and Psychology*. 2010 Sep; 25(3):367–80.
13. "Are you working too hard? A conversation with mind/body researcher Herbert Benson." Benson H. *Harvard Business Review*. 2005 Nov; 83(11):53–8, 165.
14. "The role of positivity and connectivity in the performance of business teams: A nonlinear dynamics model." Losada M, Heaphy E. *American Behavioral Scientist*. 2005 47(6), 740–65.
15. "Stirring the hearts of followers: Charismatic leadership as the transferal of affect." Erez A, Misangyi V. F., Johnson D. E., LePine M. A., Halverson K. C. *Journal of Applied Psychology*. 2008 May; 93(3):602–16.
16. "Positive affect and the complex dynamics of human flourishing." Fredrickson B. L., Losada M. F. *American Psychologist*. 2005 Oct; 60(7):678–86.
17. "The meaning of silent pauses in the initial interview." Siegman A. W. *Journal of Nervous and Mental Disease*. 1978 Sep; 166(9):642–54.

18. "Use of social-skills training in the treatment of extreme anxiety and deficient verbal skills in the job-interview setting." Hollandsworth J. G. Jr., Glazeski R. C., Dressel M. E. *Journal of Applied Behavior Analysis*. 1978 Summer; 11(2):259–69.
19. "Patient perspectives on communication with the medical team: Pilot study using the Communication Assessment Tool-Team (CAT-T)." Mercer L. M., Tanabe P, Pang P. S., Gisondi M. A., Courtney D. M., Engel K. G., Donlan S. M., Adams J. G., Makoul G. *Patient Education and Counseling*. 2008 Nov; 73(2):220–23.

Chapter 12: Compassionate Communication with Kids (pages 197–207)

1. Risley T. R., Hart B. *Meaningful Differences in the Everyday Experience of Young American Children*, 2nd edition. Brooke, 1995.
2. "An analysis of linguistic styles by inferred age in TV dramas." Lee C. H., Park J, Seo Y.S. *Psychological Reports.* 2006 Oct; 99(2):351–56.
3. "Individual differences in children's performance during an emotional Stroop task: A behavioral and electrophysiological study." Pérez-Edgar K, Fox N. A. *Brain and Cognition*. 2003 Jun; 52(1):33–51.
4. "Anxiety sensitivity, conscious awareness and selective attentional biases in children." Hunt C, Keogh E, French C. C. *Behaviour Research and Therapy*. 2007 Mar; 45(3):497–509.
5. "Temperamental contributions to children's performance in an emotion-word processing task: A behavioral and electrophysiological study. "Pérez-Edgar K, Fox N. A. *Brain and Cognition*. 2007 Oct; 65(1):22–35.
6. "Power and conflict resolution in sibling, parent-child, and spousal negotiations." Recchia H. E., Ross H. S., Vickar M. *Journal of Family Psychology*. 2010 Oct; 24(5):605–15.
7. "Problem solving, contention, and struggle: How siblings resolve a conflict of interests." Ram A, Ross H. S. *Child Development*. 2001 Nov–Dec; 72(6):1710–22.
8. "How siblings resolve their conflicts: the importance of first offers, planning, and limited opposition." Ross H, Ross M, Stein N, Trabasso T. *Child Development*. 2006 Nov-Dec; 77(6):1730–45.
9. "A school peer mediation program as a context for exploring therapeutic

jurisprudence (TJ): Can a peer mediation program inform the law?" McWilliam N. *International Journal of Law and Psychiatry*. 2010 Nov–Dec; 33(5-6):293–305.

10. We recommend *Multiple Intelligences* by Howard Gardner and *Emotional Intelligence* by Daniel Goleman as excellent introductions to these powerful and essential skills.
11. "Sibling relationship quality moderates the associations between parental interventions and siblings' independent conflict strategies and outcomes." Recchia H. E., Howe N. *Journal of Family Psychology*. 2009 Aug; 23(4):551–61.
12. "Mindful parenting decreases aggression and increases social behavior in children with developmental disabilities." Singh N. N., Lancioni G. E., Winton A. S., Singh J, Curtis W. J., Wahler R. G., McAleavey K. M. *Behavior Modification*. 2007 Nov; 31(6):749–71.
13. "Mindfulness-based parent training: strategies to lessen the grip of automaticity in families with disruptive children." Dumas J. E. *Journal of Clinical Child and Adolescent Psychology*. 2005 Dec; 34(4):779–91.
14. "A model of mindful parenting: Implications for parent-child relationships and prevention research." Duncan L. G., Coatsworth J. D., Greenberg M. T. *Clinical Child and Family Psychology Review*. 2009 Sep; 12(3):255–70.
15. "Mindful parenting in mental health care." Bögels S. M., Lehtonen A, Restifo K. *Mindfulness* (NY). 2010 Jun; 1(2):107–120. Epub 2010 May 25.
16. "Counting blessings in early adolescents: An experimental study of gratitude and subjective well-being." Froh J. J., Sefick W. J., Emmons R. A. *Journal of School Psychology*. 2008 Apr; 46(2):213–33.
17. "The effects of counting blessings on subjective well-being: A gratitude intervention in a Spanish sample." Martínez-Martí M. L., Avia M. D., Hernández-Lloreda M. J. *Spanish Journal of Psychology*. 2010 Nov; 13(2):886–96.
18. "A balanced psychology and a full life." Seligman M. E., Parks A. C., Steen T. *Philosophical Transactions of the Royal Society of London: Series B, Biological Sciences*. 2004 Sep 29; 359(1449):1379–81.
19. "Positive psychology progress: Empirical validation of interventions." Seligman M. E., Steen T. A., Park N, Peterson C. *American Psychologist*. 2005 Jul–Aug; 60(5):410–21.

20. "Journaling about stressful events: Effects of cognitive processing and emotional expression." Ullrich P. M., Lutgendorf S. K. *Annals of Behavioral Medicine*. 2002 Summer; 24(3):244–50.
21. "The effects of journaling for women with newly diagnosed breast cancer." Smith S, Anderson-Hanley C, Langrock A, Compas B. *Psycho-Oncology*. 2005 Dec; 14(12):1075–82.
22. "Writing about testing worries boosts exam performance in the classroom." Ramirez G, Beilock S. L. *Science*. 2011 Jan 14; 331(6014):211–13.
23. "The effects of counting blessings on subjective well-being: A gratitude intervention in a Spanish sample." Martínez-Martí M. L., Avia M. D., Hernández-Lloreda M. J. *Spanish Journal of Psychology*. 2010 Nov; 13(2):886-96. "Counting blessings versus burdens: an experimental investigation of gratitude and subjective well-being in daily life." Emmons R. A., McCullough M. E. *Journal of Personality and Social Psychology*. 2003 Feb; 84(2):377–89.
24. "Counting blessings in early adolescents: An experimental study of gratitude and subjective well-being." Froh J. J., Sefick W. J., Emmons R. A. *Journal of School Psychology*. 2008 Apr; 46(2):213–33. Epub 2007 May 4.
25. "Counting blessings versus burdens: an experimental investigation of gratitude and subjective well-being in daily life." Emmons R. A., McCullough M. E. *Journal of Personality and Social Psychology*. 2003 Feb; 84(2):377–89.
26. "Gratitude and subjective well-being in early adolescence: Examining gender differences." Froh J. J., Yurkewicz C, Kashdan T. B. *Journal of Adolescence*. 2009 Jun; 32(3):633–50.
27. "Reducing the racial achievement gap: A social-psychological intervention." Cohen G. L., Garcia J, Apfel N, Master A. *Science*. 2006 Sep 1; 313(5791):1307–10.
28. "Personal goals and prolonged grief disorder symptoms." Boelen P. A. *Clinical Psychology and Psychotherapy*. 2010 Dec 1. doi: 10.1002/cpp.731.

INDEX

Note: Within the index, the abbreviation *CC* refers to Compassionate Communication.

abstract concepts, 28–29, 49
acceptance-based awareness exercises, 36
affect vocalizations, 138. *See also* tone of voice
affirmations, 178–79
altruistic punishment, 85
amygdala, 24, 25–26, 27, 48
anger
and brevity in speaking, 61
effects of, 17, 174
and the "empty-chair" exercise, 175–77
ineffectiveness of, 85–86
mirroring, 97
in parent-child relationships, 204
responding to, 97, 133
sadness camouflaged by, 96
and smiling, 98
and time-outs, 97, 133, 174
and trust, 88
animals, 43, 54
anorexics, 64
anterior cingulate, 19, 48–49, 68, 100
anticipation of speakers, 82
antisocial personalities, 132
ants, 43
anxiety
and diaries, 160
effect of, on relationships, 173
exercise to reduce, 66
and mindfulness, 14, 38
and negativity, 131, 205–6
and positivity, 27–28
and rate of speech, 140
and smiling, 98
aphasia, 51
appreciation
in CC model, 123, 136–37
countering negativity with, 18
in negotiations, 188
scripted practice of, 153
attachment anxiety, 173
attention spans, 202
authenticity, 31
axons, 42

background conversations, 62
baseball, 186
bees, 54
behavioral changes, 177–78
belief systems, 81
bells, 128, 149–50
Benson, Herbert, 32, 188–89
biofeedback, 33
birds, 43
blood pressure, 126
body language, 44–45, 97. *See also* nonverbal communication
brain
  and consciousness, 54–55, 56–57
  early development of, 56–57
  effects of relaxation exercises on, 125
  and the mind, 15
  motivational centers of, 30
  resistance of, 10, 37
  See also neuroanatomy
brain-imaging technology, 38
brainstorming, 13
"breakout principle," 188–89
breathing, 33, 125, 126
brevity in speech, 15–16
  in CC model, 123, 141–42
  and conflict, 61
  and negativity, 17
  scripted practice of, 157
Broca's area, 50–51
Brooks, David, 113
business and the workplace, 183–96
  and the "breakout principle," 188–89
  brevity in, 16
  and business schools, 187
  charisma in, 183
  development of CC in, 6–9
  and facial expressions, 136
  leadership in, 185–86, 186–87, 188
  medical profession, 186, 190–94
  mirroring communication in, 83
  negativity in, 189–90, 194–96
  negotiation in, 188, 192–93
  popularity of CC in, 161
  positivity in, 186–87, 189–90
  preparing for conversations in, 192–93
  productivity in, 188–89
  three-to-one ratio in, 130
  tone of voice in, 139
  utilization of CC in, 11
  values in, 8, 110–11, 117–18, 184–86
business schools, 187

charisma, 183
children, 197–207
  and abstract concepts, 28
  and communication exercises, 161, 197–98
  dialogues with, 198–200
  effects of negativity on, 25
  parents' relationships with, 203–4
  and rate of speech, 140
  responses to suffering of, 96
  teaching communication skills to, 201–2
  writing exercises for, 204–6
clarification, 29, 84
cognitive dissonance, 82
colors, 57–58
comedians, 46
compassion
  effect of mindfulness on, 15
  effect of positive thoughts on, 35
  facial expressions of, 46
  and imagining a loved one, 99–100
  neurological basis of, 48–49
  physical effects of, 55
  and suffering, 95–96
competence, 140, 183
complainers, 31
compliments, 136, 137
comprehension
  and gestures, 44, 81
  and pauses in speech, 69
  and rate of speech, 60, 140
conflict resolution, 34, 48, 86
consciousness, 53–75

accuracy of, 87–88
in the animal kingdom, 54
boundaries of, 56
and the brain, 54–55, 56–57
definition of, 53–54
effect of mindfulness on, 15
everyday consciousness, 59–60
inner voice of, 62–69
limitations of, 59–60
of newborns, 56
and rate of speech, 140
role of, 52
speculation on, 53
constructive criticism, 180
cooperation, 77–86
among plants, 79–80
and cognitive dissonance, 82
and conflict resolution, 86
development of, 77–78
effect of, on relationships, 173
and empathy, 83–84
and expectation of reciprocation, 78
and kindness, 86
and mirroring communication, 81, 82–83
and neural resonance, 80–82
and observation, 89–90
and punishment, 85–86
social contagiousness of, 81–82
cortisol, 125, 135, 173
Crick, Francis, 54
criticism, 85, 179–80
cytokines, 179

Damásio, António, 84
dating, 166–69
Dean, Jeremy, 140
death, 113
deception
exaggeration perceived as, 31
and eye contact, 91
and facial expressions, 135
and the gaze of others, 88–90
recognition of, 48, 178
of sociopaths, 133
depression
effect of positive thoughts on, 27–28, 36
exercise to reduce, 66
and negative rumination, 25
and three-to-one ratio, 131
and writing in negative terms, 205
diaries and journaling, 160–61, 204, 205, 206
disturbing memories, 35
dolphins, 54
dopamine, 35, 98–99
Downs, Lisa J., 144
Drucker, Peter F., 184

education, 186, 187
Ekman, Paul
on breathing, 125
on expressing sadness, 95–97
on facial expressions, 44, 46, 92, 94, 95–96
on gestures, 44
on micro-expressions, 136
emotions
of children, 202
concealing emotions, 135–36
and eye contact, 90–92
and facial expressions, 91–92, 135–36, 138
inconsistency in expressions of, 95
and rate of speech, 70–74
regulation of, 174
sadness, 95–97, 127
and tone of voice, 138, 139
See also anger; fear; happiness
Emotions Revealed (Ekman), 125
empathy
effect of, on rapport, 194
and imagining a loved one, 99–100
learning, 83–84
limitations of, 84
and mindfulness, 15
and mirroring communication, 81

empathy *(cont.)*
 neurological basis of, 48
 and smiling, 98–99
 social contagiousness of, 81–82
 and tone of voice, 139
"empty-chair" exercise, 175–77
erotic words, 33–34
everyday consciousness, 59–60, 68
evolution of speech, 42
exaggeration, 31
expressive aphasia, 51
eyes
 and eliciting pleasant memories, 134
 eye contact, 89, 90–91, 134
 and fear, 94
 and honesty/deception, 88–90
 and infants, 91
 Mona Lisa smile and gaze, 99–100
 and the power of gazing, 90–94
 and smiling, 98
 and trustworthiness, 88

face
 effects of stress on, 10
 mouth, 94–95, 98–99
 and trust, 88–89
 See also eyes
facial expressions
 attending to, 135
 and concealing emotions, 135
 eliciting, 134
 emotions expressed by, 91–92, 138
 of happiness, 46
 micro-expressions, 9, 92, 135–36
 mirroring, 97
 negative expressions, 173
 neurological basis of, 44–45
 practicing, 46
 of robots, 97
 and trust, 134
 See also eyes
family. *See* loved ones
fantasies, 24–25, 27
fatigue, 18–19
fear
 and eye muscles, 94
 fearful words, 25–26
 of rejection, 173
 and smiling, 98
 and trust, 88
fight-or-flight response, 25–26
finance-related values, 112–13
first dates, 166–69
first impressions, 88, 89
Fredrickson, Barbara, 17–18, 130, 190
free association, 14
frontal lobes
 and abstract concepts, 49
 and brainstorming, 13
 and mindfulness, 38
 and negative thoughts, 26
 and neural dissonance, 173
 and positivity, 34
 and short-term memory, 15
 and stress, 10
Frootko, Cheri, 108–9

Gaia (Lovelock), 80
gazes. *See* eyes
gender, 43, 181
generosity, 38, 86
genes, 28, 32–33
gestures
 effectiveness of, 80–81
 language coordinated with, 46
 neurological basis of, 42, 44–45
goals, 81, 206
"God," perceptions of, 29
Golden Rule model of communication, 116
Goldsmith, Marshall, 184–85
Gottman, John, 130
guided imagery, 33

Hameroff, Stuart, 55
happiness
 facial expressions of, 46

and trust, 88
as universal value, 112–13
and writing exercise, 205
Hart, Betty, 201
healing, 138–39
health-care professions, 186, 190–94
heart health
effect of emotional conflict on, 179
and marital conflict, 173–74
and staying present, 126
and talking with strangers, 160
helpfulness, 78–79
honesty, 31, 88–90
hostile language, 33, 174, 179
humor, 46
hypnosis, 36

imagination
and connecting with others, 83–84
"empty-chair" exercise, 175–77
power of, 27, 37, 57
preparing for conversations, 129–30, 131
imitation. *See* mirroring
infants, 78, 91, 98–99, 100
inner speech, 63–65
constant presence of, 62–63
control of, 127–28
and intuition, 68–69
negativity in, 64
observing, 65–66
self-criticism in, 180
and staying present, 126
transforming, 66–68
vocalizing, 75
insecurity, 26–27, 180
insight/intuition, voice of, 68–69, 144–45
insula, 48–49, 68, 100
interruptions, 142–43
intimacy, 94, 127
intonations, 58
irrational beliefs, 30
irritability, 38, 66, 98

James, William, 53–54, 56
journals. *See* diaries and journaling
judgement, 85

Kanter, Rosabeth Moss, 110–11
Kaptchuk, Ted, 138–39
Kelly, Spencer, 47
kindness, 46, 86, 186
Koch, Christof, 54
Kolb, Deborah, 188

language, 10, 40–41, 125
leaders
effective communication of, 139, 188, 196
guidance from, 186–87
positivity in, 190
values of, 183, 185–86
Leahy, Robert, 36
Levin, Roger P., 117–18
life-threatening events, 113
limbic brain, 10
listening
and ability to anticipate speakers, 82
bad listening, 144
in CC model, 123, 142–43
to children, 203–4
emphasis on, 9
and inner speech, 68
and interruptions, 142–43
to monologues, 142–43, 158–59
and neural resonance, 81, 82
scripted practice of, 157
Losada, Marcial, 130, 189, 190
love, 29, 34, 89
loved ones, 165–81
and criticism, 179–80
and the "empty-chair" exercise, 175–77
first dates, 166–69
imagining, 99–100
marital conflicts, 169–73, 173–74
seeking behavioral changes in, 177–78
trust between, 173

Lovelock, James, 80
Lyubomirsky, Sonja, 30–31

major league baseball, 186
Manning, Chris, 8, 187, 193
marriage
  criticism in, 179–80
  marital conflicts, 169–73, 173–74
  word choices in, 178–79
  See also loved ones
Meaningful Differences in Everyday Experience of Young American Children (Hart and Risely), 201
mediators, 174
medical profession, 186, 190–94
meditation, 33, 36
memory and memories
  in CC model, 123, 134–35
  of children, 202
  chunking mechanism of, 59, 61
  disturbing memories, 35
  and facial expressions, 134
  rescripting negative memories, 131
  scripted practice of, 152
men, 181
mind, 15
mindfulness, 14–15, 37–38, 188–89
mirroring
  anger, 97
  facial expressions, 97
  language styles, 82–83
  and neural resonance, 80, 81
mirror neurons, 45
miscommunications, 29, 84
misunderstandings, 165–66
Mona Lisa (Leonardo da Vinci), 98
money, 112–13
monologues, 142–43, 158–59
moral decision making, 15, 89–90
mothers and infants, 98–99, 100
motivational centers of the brain, 30
mouth, 94–95, 98–99

negative rumination
  brain's perception of, 57
  and hostile language, 33
  neurological effects of, 25
  preventing, 36
  and reflecting on values, 104
negativity, 17–18
  and anxiety, 131
  and children, 202
  and consciousness, 57
  and credibility, 31
  damage caused by, 24
  effect of, on relationships, 173
  in everyday communication, 178–79
  and fatigue, 18–19
  and genes, 33
  and handedness, 44
  and inner speech, 66–68
  interrupting negativity, 26–27, 85, 129
  negative criticism, 180
  and overused words, 31
  positivity-to-negativity ratio, 18, 130–31, 189–90
  recognizing, 37
  rooting out, 131–32, 205
  self-perpetuating quality of, 25
  subliminal words, 33–34
  vocalizing, 24
  in the workplace, 189–90, 194–96
  and writing exercises, 204–5
negotiations, 188, 192–93
neocortex, 42, 57
neural dissonance, 139, 173
neural resonance, 80–82, 95–96
neuroanatomy
  amygdala, 24, 25–26, 27, 48
  anterior cingulate, 19, 48–49, 68, 100
  axons, 42
  Broca's area, 50–51
  limbic brain, 10
  mirror neurons, 45
  neocortex, 42, 57
  neurotransmitters, 35
  thalamus, 35, 57
  Wernicke's area of the brain, 33
  See also frontal lobes

neurochemistry, 24–25, 33
neuroeconomics, 78
Nexi (social robot), 97
Nhất Hạnh, Thích, 177–78
"no," the power of, 24, 30, 31
nonverbal communication
in CC model, 123, 135–36
neurological basis of, 44–45
of a robot, 97
scripted practice of, 152–53
See also facial expressions
notes, sending, 137

observation, 81, 89–90
observing self, 74–75, 145
oxytocin, 135, 174

parenting, 203–4
pausing in speech, 69–70
"peace," 35
Penrose, Roger, 55
personalities and subpersonalities, 64
Phelan, Donna, 7
Piaget, Jean, 63
pictures, thinking in, 49
plants, 79–80
positivity
in business and the workplace, 186–87, 189–90
in CC model, 123, 129–32
and children, 202
effectiveness of, 35, 36, 173
in everyday communication, 178–79
and handedness, 44
and inner speech, 66–68
in leadership, 190
neurological effects of, 34–35
and overused words, 31
positivity bias, 19
positivity-to-negativity ratio, 18, 130–31, 189–90
power of, 27–28, 28–29, 30–31
scripted practice of, 151–52
and smiling, 98
subliminal words, 33–34
writing in positive terms, 205, 206
practicing Compassionate Communication, 147–62
entering a conversation, 153–54
with a partner, 154–59
preparing to practice, 148–49
script for, 149–53
prayer, 33
preparing for conversations, 129–30, 131, 192–93
present, staying, 123, 126–27, 150–51
primates, 43
procrastination, 65
progressive muscle relaxation, 33
punishment, 85

Radin, Dean, 55–56
rate of speech
in CC model, 123, 140
and competence, 140
and comprehension, 60, 140
emotional impact of, 70–74, 138
of physicians, 138
scripted practice of, 156–57
and stress levels, 60
rational decisions, 17
reality, transforming, 34–37
reciprocation, expectation of, 78
rehearsing, 46–47
rejection, fear of, 173
relaxation
in the "breakout principle," 188–89
in CC model, 123, 124–25
effectiveness of, 27–28, 36
exercises for, 124
relaxation response, 32–33, 188
scripted practice of, 150
and staying present, 126
relevance, 62
repetitious patterns of thinking, 36–37
resistance of the brain, 10, 37
respect, 173
Risely, Todd, 201
robots, 97
Roulac, Stephen E., 194

Sachs, Joel, 79
sadness, 95–97, 127
Sartre, Jean-Paul, 89
schizophrenia, 52
The Secret (Byrne), 112
secret desires, 112
self-doubt, 173
self-esteem, 173, 185
selfishness, 77–79
Seligman, Martin, 131
short-term memory, 15–16
silence, 69–70
  in CC model, 123, 127–28
  improving, 74
  inner silence, 127–28
  scripted practice of, 151
  between words, 70–74
situational values, 113–14
slime mold, 55
slow speech. *See* rate of speech
smiles, 94–95
  and eliciting pleasant memories, 134
  and imagining a loved one, 99–100
  Mona Lisa smiles, 98, 99–100
  neurological basis of, 98–99
The Social Animal (Brooks), 113
social anxiety, 14, 91
social neuroscience, 78
sociopaths, 132–33
speed dating, 83
spiritual values, 118–19
spontaneity in dialogue, 13, 18
stress
  and the "breakout principle," 188–89
  effects of, 10, 124
  exercise to reduce, 66
  and mindfulness, 38
  and reflecting on values, 104
  and relaxation exercises, 125
  stress chemicals, 24, 135, 173, 179
  and talking with strangers, 160
subliminal words, 33–34
subpersonalities, 64
Summers, Joan, 8
suspicion, 10

tai chi, 33
taste of words, 48
thalamus, 35, 57
thinking before speaking, 18–19
Thirty Second Rule, 15–16. *See also* brevity in speaking
three-to-one ratio, 130–31, 189
time-outs, 97, 133, 174
Tolle, Eckhart, 126
tone of voice, 47–48
  in CC model, 123, 137–39
  effects of stress on, 10
  emotions conveyed by, 138, 139
  and empathy, 139
  of physicians, 138
  scripted practice of, 156
trust, 87–100
  definition of, 88
  and eye contact, 93–94
  and facial expressions, 134
  and honesty, 88–90
  importance of, 173
  and mirroring communication, 81, 97
  and smiling, 98–99
  and tone of voice, 138
twelve steps, 4, 121–45
  1. relaxing, 123, 124–25, 150
  2. staying present, 123, 126–27, 150–51
  3. cultivating inner silence, 123, 127–28, 151
  4. increasing positivity, 123, 129–32, 151–52
  5. reflecting on your deepest values, 123, 132–33, 152
  6. accessing a pleasant memory, 123, 134–35, 152
  7. observing nonverbal cues, 123, 135–36, 152–53
  8. expressing appreciation, 123, 136–37, 153
  9. speaking warmly, 123, 137–39, 156

10. speaking slowly, 123, 140, 156–57
11. speaking briefly, 123, 141–42, 157
12. listening deeply, 123, 142–43, 157

values, 103–19
in business and the workplace, 8, 110–11, 117–18, 184–86
in CC model, 123, 132–33
and conflict resolution, 86
identifying, 103–4
nature of, 111–12
in personal relationships, 115–16
scripted practice of, 152
and secret desires, 112–13
situational values, 113–14
in society, 109–11
spiritual values, 118–19
ten-day experiment, 104–9
visualization. *See* imagination
vocal inflection. *See* tone of voice

Walton, James, 115–16
warm tone of voice. *See* tone of voice
Watkins, John, 11
Wernicke's area of the brain, 33
whales, 54
White, Sara, 67–68
wisdom, 74
women, 181
workaholics, 64
worrying, 25–26
writing down thoughts
and brevity in speaking, 16, 141
in diaries, 160–61, 204, 205
writing in negative terms, 204, 205, 206
writing in positive terms, 204–5

"yes," the power of, 30–31
yoga, 33